75
70
108
114

TUSKEGEE'S HEROES

Featuring the Aviation Art of Roy LaGrone

Charlie & Ann Cooper

Motorbooks International
Publishers & Wholesalers ®

First published in 1996 by Motorbooks International Publishers & Wholesalers, 729 Prospect Avenue, PO Box 1, Osceola, WI 54020-0001 USA

Motorbooks International is a certified trademark, registered with the United States Patent Office

The information in this book is true and complete to the best of our knowledge. All recommendations are made without any guarantee on the part of the author or Publisher, who also disclaim any liability incurred in connection with the use of this data or specific details

We recognize that some words, model names and designations, for example, mentioned herein are the property of the trademark holder. We use them for identification purposes only. This is not an official publication

Motorbooks International books are also available at discounts in bulk quantity for industrial or sales-promotional use. For details write to Special Sales Manager at the Publisher's address

Library of Congress Cataloging-in-Publication Data

Cooper, Charlie.
    Tuskegee's heroes : as depicted in the aviation art of Roy E. La Grone / by Charlie and Ann Cooper : foreword, Benjamin O. Davis, Jr. : introduction to the artist, Keith Ferris : paintings, Roy E. La Grone, 1921-1993.
        p. cm.
    Includes index.
    ISBN 0-7603-0254-5
    1. United States. Air Force--Afro-Americans--History. 2. Air pilots, Military--United States--History. 3. Afro-American air pilots--Biography. I. Cooper, Ann, 1934- II. Title.
UG834.A37C66 1996
358.4'0092'273--dc20
[B]   96-2735

On the front cover: Tuskegee graduate Freddie Hutchins of class 43-D. *Roy E. La Grone*

On the frontispiece: On the flight line at Tuskegee Army Air Field: Wilson Eagleson, Harold Sawyer, Herber Houston, James Brothers, and Arnold Cisco. *Harold Sawyer*

On the title page: *These Are Our Finest*, a painting dedicated to the men of the 99th Fighter Squadron and 332nd Fighter Group. *Roy E. La Grone*

On the back cover, top: Captain Alfonso Davis, 99th Fighter Squadron, was later killed in a plane crash in Italy, 1944. *Herman "Ace" Lawson, U.S. Air Force Museum*

On the back cover, bottom: Lieutenants Andrew "Jug" Turner and Clarence "Lucky" Lester, 100th Fighter Squadron, discuss Lucky's three kills—all in one mission escorting B-17 Flying Fortresses over Germany. *U.S. Air Force Museum*

Printed in Hong Kong

Contents

In Memory of
Tuskegee Airman and Artist
Roy E. La Grone, 1921-1993

To Ester

Foreword

Tuskegee's Heroes is an accurate story of the Tuskegee Airmen's participation in pre-World War II and their later post-war important contributions to the United States. Equally important, *Tuskegee's Heroes* presents in its Chapter I, "Early Years," valuable information on black pioneers in aviation.

It is difficult to believe in 1994 that black Americans in the 1920s and 1930s were generally regarded and referred to in U.S. official military correspondence as superstitious, unintelligent, immoral, and lacking in physical courage. Yet, these beliefs were the bases for the refusal of the Army Air Corps to accept black candidates for pilot training during the 1930s. My application for pilot training upon my scheduled graduation from West Point in June 1936 was disapproved in October, 1935 by the Chief of the Army Air Corps for the stated reason that "there were no black units in the Army Air Corps and none were contemplated."

It was not until the early 1940s that the Army Air Corps was directed to create a black flying school at Tuskegee, Alabama, and train its first classes of black pilots. Simultaneously, the Army Air Corps was required to train black women and men in all of the skills required to support the creation of black flying squadrons and provide for their operational requirements.

Tuskegee's Heroes tells in interesting detail the experience of black airmen in undergoing pilot training at Tuskegee. The important role of Col. (later Brigadier General) Noel Parrish is described in some detail. The problems associated with combat in North Africa and Italy are discussed and, at the same time, the underhanded effort by the Army Air Corps to change the important tactical role of black flying units from the priority mission of tactical combat missions to the sterile mission of coastal patrol is presented.

Tuskegee's Heroes contains numerous stories of individual Tuskegee Airmen, along with the graphic presentations of their accomplishments. From these individual presentations, one can easily discern the most dramatic accomplishment—their combat achievements which led first to the immediate integration of the United States Air Force upon the issuance of President Truman's 1948 Executive Order that mandated equal opportunity and treatment for individual members of the United States Armed Forces and, later, the integration of the other military services.

Finally, one can logically proceed from the successful integration of our Armed Forces to the positive effect of this integration upon President Johnson's legislative program and the civil rights advances that followed.

Today, the Tuskegee Airmen are represented by individual Chapters nationwide. These Chapters support scholarship and other positive social programs for all Americans. They are truly America's Heroes.

Lt. Gen. Benjamin O. Davis, Jr., USAF (Ret)

During World War II, Lieutenant General Davis commanded the Tuskegee airmen of the 99th Fighter Squadron and the 332nd Fighter Group.

Preface

The compelling portraits by the talented Roy E. La Grone focus attention on a few of the valiant black American airmen who served our country so well. Mr. La Grone would have enjoyed painting them all, if time had allowed. Please accept those included as representative of the many.

In his proudest statement, Mr. La Grone said, "My life-long project… is to paint, for the Air Force Art Program, those black pilots contributing past and present." His is a rich legacy that immortalizes black aviation pioneers and the Tuskegee Airmen with whom he was so proud to serve. The authors could not have written this book without his cooperation and collaboration. It is our sincere regret that he is no longer living to enjoy the praise that we believe his work so richly deserves.

We extend our thanks to Lt. Gen. Benjamin O. Davis, Jr., USAF (Ret), and to master aviation artist Keith Ferris, Founder and Past President, American Society of Aviation Artists, for their invaluable assistance. Both knew Roy La Grone personally and held him in high esteem.

Lieutenant Colonel Alexander Jefferson, USAF (Ret), has also made a very valuable contribution. As a former POW, he has worked tirelessly to compile a list of those who spent a great deal of World War II in German stalags as Prisoners of War. Our thanks and our admiration to Lieutenant Colonel Jefferson and to all POWs.

We thank Bill Terry and all members of Tuskegee Airmen Inc., who gave of their time and encouragement in this endeavor. We extend special thanks, too, to Dave Menard of the USAF Museum's Research Division; to Floyd Thomas and staff members of the African-American Cultural Center and Museum, Wilberforce, Ohio; and to the staff of the Office of Air Force History, Maxwell Air Force Base, Alabama.

In naming the several important contributors who made this book possible, we extend our heartfelt appreciation and thanks to: Lee Archer, Roscoe Brown, Wilma and Bill Campbell, Mildred and Gene Carter, Philip and Mary Handleman, Henry Holden, Dr. Violet Jackson, Neal and Clare Loving, Charles McGee, Gertrude and Harold Sawyer, Harry J. Schaare, Maj. Gen. USAF (Ret) Lucius Theus, and Mrs. Gladys Theus, Denise Waters, Vivian White, Pat Whyte, Nancy Wright, and Cheryl Young. We are especially appreciative to Bill Adler of Adler and Robin Books, and Michael Haenggi, Commissioning Editor, Motorbooks International.

Last but not least, for her friendship and support, we are most grateful to Mrs. Ester La Grone.

Charlie and Ann Cooper

An Introduction to the Artist

Tuskegee's Heroes are fortunate to have counted among their own Roy E. La Grone, World War II pilot, Tuskegee Airman, and subsequent civilian career graphic designer, illustrator, art director, graphic coordinator, and artist. The story of the Tuskegee Airmen is better known today than it would have been without Roy's magnificent artistic portrayals of his fellow airmen and their exploits.

Roy made it a life-long project to tell the little known stories of sacrifices and contributions of the nation's black Army Air Force, and later United States Air Force, pilots through his paintings. He was uniquely positioned to do this, being driven by his own World War II Tuskegee Airman experience, his love of flying and aircraft, and his personal friendships with so many of *Tuskegee's Heroes*.

A product of Pratt Institute, Brooklyn, New York; Tuskegee Institute, Alabama; and the University of Florence in Italy, Roy was a very talented artist. His forty-four year art career began in 1949 designing and illustrating book jackets for the major New York book publishers such as Harper and Row, Charles Scribner, Random House, and the Macmillan Company.

Roy served as illustrator and designer for *Family Circle Magazine*. He created editorial layouts, mechanicals and illustrations for *AMERICA*, a full-color Russian language picture magazine distributed in the Soviet Union for the U. S. State Department.

Roy applied his talents as an Art Director for Avon Books, for Pageant Magazine, for Columbia Broadcasting System Inc., and as Art Director and Graphics Coordinator for the Medical Arts Department of UMDNJ—Robert Wood Johnson Medical School in New Jersey.

Roy's greatest artistic strengths lay in his strong sense of design and demonstration of the importance of powerful drawing with his art. He insisted on good drawing and "practiced what he preached."

The artist Roy E. La Grone as First Sergeant, Reserve Officers Training Corps, Tuskegee Institute, Alabama. *Roy E. La Grone*

A most important milestone for Roy La Grone as well as the Tuskegee Airmen has to have been his acceptance by the New York Society of Illustrators as an Artist Member in 1961. The Society of Illustrators administers the USAF Art Program in the eastern United States. It provided the vehicle for the creation of Roy's collection of Tuskegee Airmen Art.

Society of Illustrators' participation in the Air Force Art Program involves the services of professional artist members who volunteer, through the Society's Air Force Art Chairman, to participate in the Air Force mission under invitational travel orders issued by the Secretary of the Air Force. Artists travel with Air Force units, wherever they may go, for the purposes of documenting the observed mission. Time for travel, time for creation of art, transfer of ownership of the original art, and limited copyright license to the government are all donated by the artist.

Roy was very active in the Air Force Art Program. In addition to documenting the Air Force mission in the continental United States, he participated in such activities as the TEAM SPIRIT exercises in Korea and the delivery of food and medicine to locations in the dissolving Soviet Union.

In CPTP, Roy La Grone soloed a Piper Cub at Tuskegee Institute. *Roy E. La Grone*

opposite left
Roy La Grone, a student at Tuskegee Institute and member of the Civilian Pilot Training Program (CPTP). *Roy E. La Grone*

opposite right
Roy La Grone, on his final mission with the USAF Art Program, a trip to gray Siberia in a C-54 Galaxy. *Roy E. La Grone*

Roy La Grone, in the pilot-in command position of a WACO UPF 7 at Tuskegee's Moton Field. *Roy E. La Grone*

With the USAF Art Program, Roy La Grone traveled with the courtesies accorded a Colonel. He thoroughly enjoyed that role and the associated perks. *Roy E. La Grone*

Roy was long a member of the Society's Air Force Art Committee and served a term as its Chairman. The Air Force Art Collection includes 24 Roy La Grone paintings.

This book contains examples of the Air Force Roy La Grone Tuskegee Airmen collection as it stood at the time of Roy's untimely death in December of 1993. It also contains Roy's color sketches, done expressly for this book, which illustrate the stories of Tuskegee Airmen for whom Roy's planned paintings will never be completed.

Roy's art is a National Treasure cut short. He is missed by his many friends, which must include anyone who knew him. I am sure that the surviving Tuskegee Airmen feel his loss. This book provides a unique opportunity to get to know the story of the Tuskegee Airmen while also getting to know an excellent artist who dedicated his life not only to his art, but to the memory of his fellow black pilot friends.

I am proud to have served with Roy throughout his years as a member of The Society of Illustrators and know that he will be with us as we jury future year's Air Force Art submissions.

Keith Ferris
Morris Plains, New Jersey

Keith Ferris is a Founder, Past President, and Artist Fellow of the American Society of Aviation Artists (ASAA), an Honorary Vice President of the British Guild of Aviation Artists (GAvA), and a member of the U.S. Air Force Art Program and the Society of Illustrators.

CHAPTER ONE

Early Years

Turmoil followed the end of the Civil War in the United States, despite an idealistic restoration program that purportedly addressed civil rights. More than 186,000 blacks served in that war and 38,000 gave their lives, yet their heroism went unheralded. Virulent racism existed in the nation. Just after that war, the 9th and 10th Cavalry were created as an experiment—not the last "experiment" created due to shortages of manpower and untapped capabilities of black men and women. Called the *Buffalo Soldiers*, men of the 9th and 10th Regiments fought Native Americans on the Great Plains, in Arizona, and in Texas. A monument to these forerunners of the later Tuskegee Experiment is displayed at Tuskegee Airman National Museum, Detroit, Michigan.

Major General Lucius Theus, USAF (Ret), said of the monument, "General Colin Powell was our keynote speaker for the dedication of this particular sculpture. Since many of our Tuskegee Airmen did serve in the armed services prior to World War II—some of them as *Buffalo Soldiers*— we feel very much a part of them.

"In the late 1800s, many trained combat persons wanted to remain in the service. Transferred to the west, the majority of those soldiers who protected the settlers and won the west were blacks. Native Americans named them because their woolly hair resembled the manes of buffaloes. The name was really a sign of respect and fear as they were regarded as great warriors. They performed many acts of bravery, some of which are recorded and some of which are not."

During the decade of the 1930s, the two regiments were posted in Kansas, Virginia, and New York. By 1938, some talented and irrepressible blacks in addition to the *Buffalo Soldiers* had also gained token gestures of recognition and respect, having been less than a year since Amherst College graduate William H.

Hastie was confirmed as the first black federal judge. There were further gains in 1938: an honorary doctorate bestowed upon opera star Marian Anderson, an unprecedented legislative seat won by Crystal Bird Fauset in Pennsylvania, and Richard Wright's *Uncle Tom's Children* achieved best-seller status.

Yet, 1938 was contradictory.

Domestically, citizens of the United States struggled to overcome the economic devastation of the Great Depression. On a global scale, Americans vacillated between feelings of guilt that British citizens were being issued gas masks, for example, and desperate yearnings for peace and isolation from the war that threatened to rage like a wildfire out of control.

Willa Brown

As the flames of that war—a second World War—threatened to spread across the Atlantic, the Civil Aeronautics Authority (CAA) launched the Civilian Pilot Training Program (CPTP) in September 1939. The CAA, a non-military governmental organization, sought to increase a potential military pilot reserve. The program effectively threw open wings of opportunity—aviation flight training—to qualified civilians.

TIME magazine reported, in 1939, "It [the CAA] certified 220 U.S. colleges and universities for participation in its pilot training program, prepared to name still more to share $5,675,000 voted by Congress for schooling 11,000 new fliers this year. The trainees are all civilians, most are collegians. They will be taught to fly by commercial air schools, at a cost to the United States of $290 to $310 per student. When they graduate, they will be far from qualified as military pilots, but most of them should rate private pilots' licenses (allowing them to fly themselves and passengers for fun, do no flying for hire). But CAA's fledglings, with the rudiments of flying,

---

"Lynch 'im!" The cry struck terror into the heart of a child. Eugene Jacques Bullard, born October 4, 1894, cringed into the corner of his family's small cabin in Georgia when his father, a native of the French colony in Martinique, was threatened by the hooded Klansmen that paraded not far away. Eugene's mother had died when he was six and, with a bravery beyond his years, Eugene ran away to escape the terror of prejudice and bigotry when only eight years old! He did, indeed, escape. His brother was hanged later by a lynch mob.

"I will somehow find my way to France," the youngster vowed, having heard of the land from his father, who praised it as a country in which a black could live without fear for his life.

Incredibly, the runaway child survived. He scrabbled for food to eat and a place to work and sleep. Among other things, he tended horses for Gypsies and rode as a jockey. At age ten, he miraculously hitched rides on railroad boxcars like a seasoned hobo and discovered the Atlantic coast at Norfolk, Virginia. He found a steamer, loaded and ready to sail, on which he stowed away.

Eugene didn't know how to find France. He didn't know the destination of his ship. He simply knew that he was searching for a land where blacks received fair treatment. His was a dogged determination.

Put off the ship in Aberdeen, Scotland, Bullard grew from ten to seventeen as a loner in the United Kingdom. He learned to prize fight and, eventually, found his way to the Paris of his dreams. He arrived there as a welterweight having boxed professionally in England and North Africa.

At the onset of World War I, Bullard enlisted in the French Foreign Legion. Transferred to the 170th Infantry, he was wounded, hospitalized, and returned to service more than once. Having been declared disabled by the infantry, he re-enlisted, this time in the French Air Force. Trained as a military pilot—unheard of at the time for a black man—Eugene Jacques Bullard made history. The first of his race to pilot a military aircraft, he proved to be as scrappy in aerial combat as he was in the boxing ring.

By the end of the war, the honors bestowed upon Bullard included the Chevalier of the Legion of Honor, the Croix de Guerre, the Cross of the Lafayette Flying Corps, and the Medaille de Verdun, and eleven others—all awarded by his adopted land. His accomplishments were virtually ignored in the country of his birth. The U.S. government never officially recognized the contributions of this daring "Black Swallow of Death," who proudly demonstrated for the world that which he had painted on the fuselage of his aircraft—"all blood runs red."

This illustration of Eugene Jacques Bullard was painted by Harry J. Schaare, a colleague of Roy La Grone, fellow member of the New York Society of Illustrators and the Air Force Art Program. Against a Nieuport 17, La Grone modeled for this work, created for a magazine article.

Schaare, currently of Sedona, Arizona, flew C-47s during World War II. He enlisted in 1942, entered aviation cadet training in 1943, graduated in 1944 and instructed in AT-10s prior to being sent overseas. He said, "When I got back from overseas, I was based at Fort Benning, Georgia. Multi-engine rated on my civilian pilot license, I searched for a place to fly to accrue ten hours in a single engine aircraft to get my rating. Tuskegee Army Air Field was the closest. At the Flight Office at TAAF, the instructor said, 'Let's go,' and headed toward an AT-6 on the line. I had no helmet, no goggles. I looked around at group of Tuskegee Airmen standing around. One of them tossed me a pair of goggles and a helmet. He said, 'Here you go.' The Tuskegee Airmen were great. In no time I had the required ten hours and qualified for the rating on the license. There is a special camaraderie among pilots. We used to call non-pilots, Ground Pounders, Paddle Feet, Gravel Agitators, Grunts.

"There was an attitude about being a pilot."

GALVESTON ✈ NEW YORK

WORLD'S FAIR

SUPPORT

AUGUST 1940

AFRO-AMERICAN
Air Derby

Willa Brown and Cornelius Coffey put their energies and talents into the promotion of an all-black Air Derby in August 1940. *Floyd Thomas and African-American Cultural Center and Museum, Wilberforce, Ohio*

will be far better material for the Army and Navy air corps than total greenhorns."

One civilian flier, the article continued, who eagerly anticipated the CAA's announcement and sought to increase the numbers of black pilots in the Chicago area was Willa Beatrice Brown, then aged 29. There was no chance that Willa Brown would *fly* for the military, she had two strikes against her—her race and her gender. (There is bitter irony in the fact that women were found qualified to *teach* male pilots for World Wars I and II, but were not deemed qualified to fly combat for their country until 1993!)

Willa Brown was one more of the capable female pilots who was, in this instance, born too soon. But, the *TIME* article acknowledged that, "in her role as Secretary of the National (Negro) Airmen's Association and as one of the few black women pilots holding a limited commercial certificate, she did work avidly to interest fellow blacks in flying, to help obtain for them a share in the CAA's training program."

At that time, Willa Brown ran Brown's Lunch Room at Harlem Airport near Chicago, a full partner with her husband, Cornelius R. Coffey, in their flying service. Willa, a true aviation pioneer, represents all educated, ambitious, enthusiastic, and determined U.S. airmen and airwomen. Trained at Indiana State Teachers College, this educator exchanged a blackboard and school room for a cockpit and airport at Aeronautical University in Chicago's South Loop in 1939. She made lasting and valuable contributions for women, for blacks, and for aviation.

After a course in aerobatics in 1939, Charles Alfred "Chief" Anderson picked up an airplane that had been purchased by and was intended for Tuskegee Institute. "I flew it to Tuskegee," he recalled, "becoming the first black pilot employed by the school." By that time, Chief had logged 3,500 hours of flying.

"During a visit to Tuskegee Institute," he said, "Mrs. Eleanor Roosevelt, the wife of U.S. President Franklin D. Roosevelt, asked me to take her for a flight over the area. Despite the tremendous opposition of those with her—Secret Service men and the like—Mrs. Roosevelt was willing to risk her life with one of us because she saw no reason why blacks could not pilot airplanes."

At a time when few people had faith in the ability of blacks to fly, Mrs. Roosevelt was apparently influential in convincing President Roosevelt to further support the black pilot training program at Tuskegee. "Shortly thereafter," said Chief Anderson, "Tuskegee Institute was selected to participate in a program with the U.S. Army Air Corps to find out if blacks could measure up as military pilots. Their records speak for themselves."

Roughly 1,000 pilots who successfully completed Army Air Corps Training were under the tutelage of Chief Anderson in the primary phase. He went on to serve as Director of the Negro Airmen International (NAI) Summer Flying Camp at Tuskegee University and to teach ROTC cadets. He admitted, "I cannot say how many individuals I have taught to fly nor can I say how many hours I have spent in the air with this objective."

Born in Bridgeport, Pennsylvania, in 1907, Chief Anderson attended Drexel Institute in Philadelphia, the Chicago School of Aeronautics, and the Boston School of Aeronautics. He was honored at the Gathering of Eagles, Maxwell Air Force Base, Montgomery, Alabama, in 1990 and, often dubbed the Father of Black Aviation, was awarded an honorary degree during the 104th Spring Commencement Exercises of Tuskegee University: the degree was presented to Charles Alfred Chief Anderson, Doctor of Science. Chief Anderson, beloved pioneer airman, came to the end of his auspicious live in the spring of 1996.

Willa was working on a Master's degree at Northwestern University, Illinois, when flying became her passion. She earned her pilot's certificate in 1939 at a crucial time in history. U.S. President Franklin Roosevelt was requesting more than $500 million for defense and demanding of Hitler and Mussolini assurances that they would not attack America. War's flames licked at our nation like poised rattlesnakes' tongues and United States industry began gearing up to meet them. Pilot training was being made available for thousands and Willa understood the ramifications of such an opportunity. She set her sights on a certifi-

Dr. Albert Forsythe was born and raised in the Bahamas. Having attended Tuskegee Institute, he graduated from dental school at McGill University, Canada, and opened a practice in Atlantic City, New Jersey.

Despite society's attitudes, Forsythe purchased an airplane and obtained flight instruction from Ernie Buehl and Chief Anderson. His pilot's certificate made him proud of the double honor of a license to fly and of demonstrating further proof of the capabilities of blacks.

He planned long-distance flights with Chief Anderson, one in the Fairchild 24 flown to Los Angeles and back by the pair in July 1933. The only instruments on board were a compass and an altimeter. They experienced severe rain, hail, strong winds, and turbulence, yet they successfully completed the first round-trip transcontinental flight flown by blacks. Upon their return, they were cheered by more than 15,000 onlookers while guests of honor in a parade in Newark, New Jersey.

Forsythe and Anderson flew to Canada the same year—the first blacks to fly across an international border.

Their third and much-heralded flight, to Forsythe's native Nassau, was flown in 1934. Like Lewis and Clark, the pair departed for uncharted territory, island-hopping more than 12,000 miles through the West Indies to South America. This ambitious flight attracted world-wide attention and the pair, though struggling for recognition from their own country, were often greeted warmly by presidents or governors at their destinations.

Dr. Forsythe practiced dentistry in Atlantic City and in Newark. In 1985, he was elected to the Aviation Hall of Fame of New Jersey. February 7, 1986 was proclaimed "Albert Forsythe Day" by the Mayor of Newark. Memorabilia of his auspicious aviation career is exhibited at New Jersey's Teterboro Airport Aviation Museum.

cated flight instructor (CFI) rating, determined to teach flying to as many blacks as possible at a school certified by the CAA's program.

TIME's article noted that Willa was secretary of the NAAA (National Airmen's Association of America), one of the first black aviation associations, but failed to mention that this beautiful and vivacious woman participated in its creation. Coffey and his wife opened the first flight school owned and operated by blacks. Her role, once she and Coffey had prevailed and the school was authorized to train pilots under CPTP in 1940, was as

director and coordinator of training at that Coffey School of Aeronautics. She held that position until 1945, establishing a "first" as the first black woman to serve in the Civil Air Patrol.

Among Her Mentors: Bullard and Coleman

Willa Brown must have read accounts of those who had learned to fly when airplanes were flimsy wooden and rag machines. She must have thrilled to tales of aerial exploits of a time when pilots flew more on faith than on knowledge. She could not have failed to have read of pioneering blacks, the

19

first to fly in this country and in the world. She would have eagerly devoured every story available about Eugene Jacques Bullard in France and, closer to home, of Bessie Coleman in Chicago. She must have felt inspired by their courage; two who ignored racist barriers and forged their own way, instilling admiration and hope.

Eugene Bullard

Eugene Jacques Bullard, born in Georgia on 4 October 1894, was the son of a native of Martinique. Told by his father of "a land across the sea in which blacks were treated fairly," a precocious Bullard ran away from home as a child. Only eight years old, but with a fervor that belied his years, the young Bullard was determined to find France. Without even knowing about the Atlantic Ocean that separated him from his concept of freedom, he worked and questioned until he found Norfolk, Virginia. Eugene stowed away on a steamer, reaching Aberdeen, Scotland, before being put off the ship.

The enterprising Bullard learned to box and to support himself with his fists. At the start of World War I, he enlisted in the French Foreign Legion. After being wounded while serving in the infantry, Bullard re-enlisted, this time in the French Air Force. Trained as a military pilot—a first for the world!—he was honored for his combat flying, but only by France, never by the land of his birth.

The youthful Bullard was searching for a land where he believed that blacks received fair treatment. His was a dogged determination.

As a loner in the United Kingdom, Bullard grew to his late teen years. He learned to prize fight and, eventually, found his way to the Paris of his dreams. He arrived there as a welterweight having boxed professionally in England and North Africa.

Transferred to the 170th Infantry, he was wounded, hospitalized, and returned to service more than once. Having been declared disabled by the infantry, he re-enlisted, this time in the French Air Force. Flying as a combat pilot, Eugene Jacques Bullard made history.

Bessie Coleman

Bessie Coleman, born in 1893, a decade before the first flight of the Wright Brothers, was sixteen when Harriet Quimby became the first woman licensed to fly in the United States. Like Bullard, she sought France, solely to find someone willing to teach her to fly. Denied by flight schools in the United States, she had two major obstacles—she was a black and a woman!

"Civilian Pilot Training—1940"

Born February 1, 1921, in Pine Bluff, Arkansas, Roy La Grone served in Senior Army ROTC, Tuskegee Institute, with the rank of First Sergeant. He completed Civil Pilot Training Program at Tuskegee in 1942. Drafted into the Army Air Corps as a Sergeant, La Grone was sent to Camp Robinson, Little Rock, Arkansas, and subsequently assigned to Chico Army Air Corps Flying School, Chico, CA. Upon completing training at Chico, he was assigned to Army Air Corps Administration School in Fort Logan, Colorado, from which he graduated March 1943. La Grone was assigned to Tuskegee Air Corps Flying School and served with Special Services, 318th Air Base Squadron, commissioned a Flight Officer in 1943. Transferred to Allied Military Headquarters, Caeserta, Italy, in 1944, La Grone studied art at the University of Florence, Italy, before returning to the United States.

La Grone was honorably discharged from the Army Air Corps in January of 1946. In his proudest statement, this creative artist said, "My life-long project… is to paint, for the Air Force Art Program, those black pilots contributing past and present." His is a rich legacy that immortalizes the Tuskegee Airmen with which he was so proud to serve.

Bessie spent her hard-earned money to travel to Europe in hopes of finding flight training. She victoriously received Fédération Aeronautique Internationale License #18,310 in June 1921—the first black to be licensed to fly.

Years later, the Soldier's Medal was pinned on Sgt. Arthur Freeman of the Tuskegee Army Flying School by Maj. Gen. Ralph Royce, Commander, Army Air Forces Southeast Training Center. Arthur Freeman, Bessie's nephew, was the only one of her family to follow her into aviation, despite the fact that Bessie's admirable goal had been to bring flying to as many blacks as possible.

Elois Patterson, one of Bessie's three sisters, wrote an article about *Brave Bessie*, from which the following is excerpted:

opposite
"Brave Bessie" Coleman, first black in the world to be licensed as a pilot, received her training in France where her race and gender were not deterrents. *U.S. Air Force Museum*

Bessie Coleman was called *Brave Bessie* because she had fearlessly taken to the air when aviation was a greater risk than it is today and when few men had been able to muster such courage. An avid reader, Bessie was well informed on what the Negro was doing and what he had done. Given the opportunity, she knew he could become as efficient in aviation as anyone. She toyed with the idea of learning to fly, even displayed an airplane made by a Negro boy in the window of the barber shop in which she was a manicurist.

After having been refused repeatedly at each aviation school to which she applied, sometimes because of her race and sometimes because she was both a Negro and a woman, she took her quest to the late Robert S. Abbott, a founder, editor and publisher of the *Chicago Weekly Defender*. He advised her to study French and Bessie promptly enrolled in a language school in Chicago's Loop. That accomplished, he assisted her in contacting an accredited aviation school in France. She planned to obtain certification and return to the United States to open an aviation training school for young blacks.

Bessie made two trips to Europe, returning to Chicago from the second one in 1922. Holder of a certificate from the FAI, she enlisted the managerial assistance of David Behncke, Founder and President of the Air Line Pilots Association (ALPA). She put on an air exhibition in 1922 at Checkerboard Field, today known as Midway Airport, Chicago, after which she received many calls from young Negro men, anxious to learn to fly. Bessie had obtained her certificate at great personal expense and sacrifice. She told prospective students that they had to wait until either some forward-thinking blacks opened a training school or until Bessie herself could give enough demonstrations and accrue sufficient money to undertake opening a school herself.

Bessie Coleman. For whom William J. Powell, Los Angeles, California, initiated aero clubs in honor of. *U.S. Air Force Museum*

"...I was in Operations a long time. I learned that you had to go out of your way to show interest and be sympathetic and understanding and to give credit to people."

That statement epitomized the remarkable Commander, Tuskegee Army Air Field, Brig. Gen. Noel F. Parrish. Born in Lexington, Kentucky, in 1909, the son of a Christian minister, Noel Parrish was but sixteen when he matriculated at Rice Institute, Houston, Texas. He graduated at age nineteen and re-entered Rice to take graduate courses in 1929. A victim of the Great Depression, Parrish left Texas for California to avoid the embarrassment of being without a job. Approached by U.S. Army recruiters, he obtained his father's permission (He was only twenty!) to enter military service. After a year in the cavalry, he applied for and was accepted for flying training.

Sent to March Field in 1930, Parrish completed the four months of primary training in the last class to be taught at March. A member of the first class to obtain basic training at Randolph Field, Texas, he graduated in 1932, having also trained in aerial combat at Kelly Field. "When finished, you were commissioned a 2nd Lieutenant, Reserve," he said, "After one year of active duty, you were out."

Discharged in 1933, the Depression at its height, Parrish faced a choice. He said, "The one thing you could do, which was still considered somewhat disgraceful, was to enlist. If you enlisted as a pilot, there you were... We flew airplanes for depots, transport flights. We enlisted pilots were titled Air Mechanics."

Taken on at Chanute Field, Illinois, as a private first class, he was then transferred to the depot at Dayton, Ohio. After the presidential election of Franklin D. Roosevelt, openings were created for thirty-six positions in the Regular Air Corps. In 1935, out of more than 200 applicants, Parrish was number six of the 36 taken back into the Corps as an officer and he was returned to the 13th Attack Squadron.

Parrish was sent to Glenview, Illinois, as assistant supervisor of a contract flying school. Charged with teaching civilian instructors to teach according to military standards, he also flew with students to ensure those standards were met. A bill was passed that ensured that two schools would be for black pilots: at Tuskegee and, with Brown and Coffey, in Chicago. Parrish's involvement with Coffey's School of Aeronautics led directly to his assignment as Air Force supervisor in primary flying at Tuskegee.

Brigadier General Noel F. Parrish and the Members of 42-C, the first Graduating Class

Mac Ross
Charles De Bow
Lemuel Custis
Benjamin Davis, Jr.
George Roberts

"This was a big job—getting started in black pilot training and to make it a success, to at least offer an opportunity for some eligible blacks and to furnish an example. Actually, it turned out to be an inspiration to many. …It became necessary to know and to understand so much about the feelings and attitudes of individuals involved, both civilian and military at higher levels, and to know how to get decisions and to influence opinion. I couldn't honestly say that I thought the thing would run properly if I got out."

Parrish started as supervisor of the primary school at Tuskegee. When the first class was sent to Tuskegee Army Air Field for basic training, he was moved to be director of training at TAAF, then promoted to Commander. He remained until mid-1946. Parrish became successful as a writer, submitting articles under the pen name, "Frank Lambert," so that, in his words, "I wouldn't have to go through clearance headaches. I wrote several articles about [my] experiences, other pilots' experiences, and on instruction techniques." Later, when the commander at Tuskegee was unsympathetic toward blacks, Parrish wrote an article that was understanding and compassionate. He said, "It was during the war. I was the Director of Training and managed to get all of my views in the article. This got a certain amount of play in the press. The commander never guessed that his own chief of training had written the article."

To attest to the esteem in which he was held, Brigadier General Parrish was honored with a testimonial, sponsored by the East Coast Chapter at Andrews Air Force Base, September 1986.

Brig. Gen. Noel F. Parrish:Tuskegee Airmen, products of your tireless efforts in a domestic environment as hostile as in any war-torn world, express our heartfelt gratitude for your:

> *Flawless and fearless leadership*
> *Rare faith and confidence in our potential*
> *Independence in thought and aspirations*
> *Endless courage and patience*
> *Never-ending demand for excellence*
> *Dedication to the principle of impartiality*
> *Sensitivity to our diverse and special needs*
> *Honesty in dealing with us and for us*
> *Integrity in the face of monumental opposition*
> *Personal devotion to the unity of humanity*
> *At great risk to your military career.*
> *These qualities were inspirational and instrumental in our success and final victory in the "Tuskegee Experiment," designed for failure. With most sincere and respectful appreciation, we, your fellow-airmen and friends, salute you.*
> *May God bless you and keep you."*

Col. Harry Sheppard, USAF Retired

Cornelius Coffey became the first black Airframe and Engine mechanic in 1931. With his wife, Willa Brown, he trained 1,500 fledgling military pilots. *Floyd Thomas and African-American Cultural Center and Museum, Wilberforce, Ohio*

front entrance, although they were separated once inside the grounds.

…She decided to make an all-out effort to establish a school where she could train young Negro men to fly.

I remember one letter she wrote me saying she had taken an escort, and even went to a pool room, so determined was she to have Negro men become air minded. The very last letter that I received from her said, 'I am right on the threshold of opening a school.'

…but she lived to do no more. On the night of April 30, 1926, at 7:30, Brave Bessie was killed.

In an accident that was not unlike that in which the first U.S. licensed woman pilot, Harriet Quimby, was killed in 1912, Bessie was catapulted out of a tumbling aircraft. She had worn no safety belt!

Inspired by "Brave Bessie," William J. Powell, another of the nation's black pioneers, organized Bessie Coleman Aero Clubs to promote flying among blacks. In 1931, his flying club sponsored the first all-black air show in the United States and, through the Bessie Coleman School, contributed to a dramatic growth of blacks in aviation. Powell published *BLACK WINGS* in 1934, dedicated to Bessie. In it he urged black young people to carve their own destiny—to become pilots, designers, and business leaders in the field of aviation. He exhorted men and women to "fill the air with black wings."

Among Her Contemporaries: Coffey, Bragg, and Spencer

Cornelius R. Coffey

Born in Arkansas in 1903, the year that the Wright Brothers introduced powered flight, Cornelius Coffey moved with his family to Chicago in 1923. A start in aviation was difficult for him in the early 1920s. But, Coffey said, "When you set your mind to do something, you can usually find a way—if you try hard enough."

Attending Automotive Engineering School, he took a job as an auto mechanic after graduation. "At that time," said Coffey, "I could see that there was not much difference between automobile and aircraft engines, so I decided to use the mechanical training and convert to aircraft, figuring that some day there would probably be quite a demand for it."

In 1928, he made his first solo flight in a Waco 9 that he had to purchase—a black could find no flight schools from which aircraft could be rented—and he

Bessie barnstormed across the country and undertook a rigorous program of speaking engagements. The late Ross D. Brown said about Bessie, in his book *Watching My Race Go By*, "She fought and flew ahead of her time and was greatly misunderstood."

When Bessie appeared over the town in which she was reared, Waxahachie, Texas, she was permitted to use the university grounds of the whites for her exhibition flying. She refused to exhibit unless her people were allowed into the grounds through the

Lemuel Rodney Custis
Class 42-C

A graduate of the first class to complete Aviation Cadet Training at Tuskegee, Lemuel Custis was one of five out of an original thirteen to be granted the coveted wings of the Army Air Corps and to attain the rank of 2nd Lieutenant. During World War II, he participated in the European/Mediterranean Theatre as a fighter pilot and as Squadron Operations Officer, serving in Africa, Sicily, and Italy, credited with one confirmed aerial combat "kill" and two "probables."

Custis was educated in Hartford, Connecticut, public schools and graduated from Howard University in 1938. Accepted into the Army Air Corps for Aviation Cadet Training at Tuskegee, AL, he graduated in March 1942 as a 2nd Lieutenant and Pilot. He was assigned to be one to fly in the first combat mission on Pantelleria, but, according to Charles Francis, "his aircraft developed engine trouble en route from OuedN'ja in North Africa." In subsequent air combat, Custis shot down a FW-190. He was quoted as having said, "It was the roughest day of the campaign, but the hunting was good all day."

Married to Ione Williams, Custis attended the University of Connecticut School of Law and joined the Tax Department of the State of Connecticut in which he served for over thirty years. In 1975, became the first black to attain a managerial position as Chief of Sales Tax for the Tax Department.

Another of Tuskegee's Heroes, Custis was awarded the Distinguished Flying Cross, Air Medal with Oak Leaf Clusters, and campaign medals from African, Sicilian, and Italian campaigns. In addition to serving on several commissions, Custis has been advisor to the Connecticut Aeronautical Historical Association, Bradley Air Museum, Windsor Locks, CT.

never parted with that biplane throughout his lifetime. Enrolled in Curtiss Wright Flying Service, he completed a Master Mechanic Course for Airframe and Engine (A&E); in 1931, he became the nation's first black man to hold an A&E certificate.

Coffey enrolled in the Curtiss Wright Flying Service. That statement is too simplified. According to Rufus Hunt, "Coffey and John C. Robinson, two young automobile mechanics, had been trying unsuccessfully to enroll in an approved aircraft and engine mechanics' school, denied admission because of their race. When an advertisement for students to enroll in the fall classes of the Curtiss Wright School of Aeronautics appeared in the aviation magazine, *Aero Digest*, Coffey answered it without mentioning his race. He and John C. Robinson were accepted and all summer they made payments toward their tuition via federal money orders. When their color was discovered, the school tried to return their money. At the time, Coffey worked as an auto mechanic for Mr. Emil Mack, Elmwood Park Motor Sales. Mr. Mack advised them to refuse to accept the money and stated that he would support them in a lawsuit against Curtiss Wright if they were refused admission to classes. Under this threat, the school relented and, despite continuous harassment by their white classmates until the teacher intervened, the two men persevered and finished the course at the top of the class.

"…On graduation, Coffey and Robinson were told that the school's doors would never again be closed to blacks and that 'if we could get a large enough group together to make up a class, we'd be employed as assistant instructors,' Mr. Coffey recalled."

Approximately thirty students, friends of Robinson and Coffey, signed up for the course.

In 1931, he became involved with the establishment of the black-owned airport at Robbins, Illinois; then moved to Harlem Airport after a devastating storm wrecked the building and aircraft at Robbins. With four J-3 Cubs and a flight instructor certificate, Coffey started his own school. With the CPTP contract, he upgraded to a fleet of Waco UPF-7 biplanes and began training pilots for the military. Approximately 1,500 black pilots have Coffey and Willa Brown to thank for their training.

Rufus Hunt wrote, in his monograph, *The Coffey Intersection*, "When he was asked if he harbored any ill feelings for the system that imposed so much adversity upon him, Coffey's answer symbolized the plight of his people: Segregation forced him to work harder to achieve and overcome, for this reason he became stronger.

Coffey devised methods of cold weather operation for aircraft and, according to Hunt, "…created the carburetor heat control used on today's modern airplanes.

Chauncey Spencer, left, and Dale White piloted "Old Faithful" on a well-publicized cross country goodwill flight aimed at attracting national attention to the plight of black aviator. From the *Pittsburgh Courier*, May 12, 1939. *Chauncey Spencer and U.S. Air Force Museum*

INTREPID CHICAGO AVIATORS PILOT 'OLD FAITHFUL' IN CROSS COUNTRY GOODWILL TOUR

George Spencer "Spanky" Roberts
Class 42-C

George Spencer "Spanky" Roberts, first to command the 99th immediately after having received his wings, graduated from West Virginia State College. He served as Operations Officer in the original 99th Fighter Squadron as it entered the war and was assigned to North Africa.

Spanky flew with the squadron at Anzio, January 1944, surviving a dogfight—twelve pilots outnumbered two-to-one—with a gaping hole in his left wing. The 99th pilots downed five enemy planes and, later, three more. Spanky remarked, "This is the first time in five months that we encountered enemy opposition. In less than five minutes. …We poured hell into them."

Spanky, married to Edith McMillan, took over as Commander of the 99th from Col. Benjamin O. Davis, Jr., in September of 1943. An excellent pilot, he led as many as 64 planes as cover for heavy bombers and on strafing missions.

First serving as Deputy, he became Commander of the 332nd Fighter Group from 1944 to war's end. Returning to Tuskegee Institute, Spanky became Professor of Air Science and Tactics. In 1950, he became the first black commander of any integrated USAF unit, assigned to Langley Air Force Base, Virginia. He graduated from Army Command and General Staff College, Fort Leavenworth, Kansas, where he and his family lived on the Fort as first blacks in 125 years of its history.

Spanky served at Suwan Air Force Base, Korea, Chanute Air Force Base, Okinawa, Griffiss Air Force Base, NY, and as Director of Support for all fighter planes in Vietnam, and the F-104 program for 18 Allied Nations and Ground Radar, McClellan Air Force Base, CA.

Spanky was a Command Pilot with more than 6,000 flying hours. He had been decorated with the Distinguished Flying Cross, Air Medal, and seven Commendation Medals each with three to nine Oak Leaf Clusters.

In 1944, he spoke in Cleveland, Ohio, as a tree was planted in memory of Lt. Sidney Brooks. "He remarked, 'Sidney Brooks fought and stood for the things that are America. America is its people—you. We are finding that the color of a man's skin, the blood in his veins, or his religion makes no difference in finding whether he is or is not a man. The privileges of America belong to those brave and strong enough to fight for them.'"

Spanky spoke from experience, and from the heart.

Colonel George "Spanky" Roberts, USAF Retired, died at the age of 65. Brig. Gen. Joel M. McKean said of him, "It is appropriate that we honor Colonel Roberts, for indeed he was a distinguished American, an Air Force leader, dedicated husband and father. He was a vital link in America's maturity. Spanky Roberts proved to us that the realization of a dream is independent of race, color, or creed."

"…Recalling the four days he spent repairing the broken crankshaft for the engine of his four-place Robin Challenger while barnstorming throughout the South in 1937, Coffey noted that a crowd of white people gathered to jeer and tease. He said, "They wouldn't allow the colored to stop and watch, they'd run them away from that fence. They didn't want to give them no bad ideas."

In September 1979, Coffey was honored at the National Air & Space Museum, Smithsonian, along with Dr. Lewis A. Jackson, C. Alfred "Chief" Anderson, Dr. Albert E. Forsythe, and Willa Brown. An air navigation fix was named in his honor in 1980 and 22 July was proclaimed "Cornelius Coffey Day" in Chicago.

Coffey, one of the outstanding vanguard of black pilots who shared a passion for the air with all who followed, never lost that love. "Flying," he said, "is hard to give up, because up there in the skies you never know from one minute to the next which part of your wit and energy you are going to have to draw on."

It is significant that a white man who would become one of the most beloved by the Tuskegee Airmen—Brig. Gen. Noel Parrish—proved a direct link between Willa Brown, Cornelius Coffey, and Tuskegee Army Air Field. Parrish was sent to Glenview, Illinois, as assistant supervisor of a contract flying school. Charged with teaching civilian flight instructors to teach according to military standards, he also flew with students to ensure that those standards were maintained. Since two schools were for black pilots, Parrish became involved with Coffey's School of Aeronautics. This experience led directly to his assignment as the Air Force supervisor at the primary flying school at Tuskegee.

Janet Waterford Bragg

Chicago was a beehive of black aviation activity during the 1930s. Many learned to fly despite barriers to their progress.

Charles H. DeBow, Jr. Class 42-C

Born in Indiana and a holder of a private pilot license through civilian training, Charles H. DeBow, Jr., was eagerly sought as proof for dubious officers of the Army Air Corps that blacks could become flying officers. An Indiana attorney, Willard Ransom, sent DeBow current War Department releases. Ransom wrote, "…Will you inform me as to whether or not you are interested in (flying for the military) or whether you have been contacted as yet by the Army Officials?"

A letter from Walter White, Secretary, on the official stationery of the National Association for the Advancement of Colored People was sent to DeBow in 1941. The letter stated, in part, "Dear Mr. DeBow: Will you write me … if you would be willing to join the Air Corps of the U.S. Army if given an opportunity and if you come within the general requirements as to age, freedom from dependents, and other qualifications given in the enclosed memorandum? We are writing to each of the colored men who have pilots' licenses from the U.S. Department of Commerce.…"

Although DeBow had typhoid as a child and although his difficulty with color recognition on eye exams threatened to keep him from becoming a pilot, he overcame all challenges. DeBow graduated with the first class at Tuskegee Army Air Field.

Newly commissioned a Lieutenant, DeBow was asked by a white civilian in Alabama, "What do you boys want to fly for, anyhow?" Concerned that he was unable to give what he felt was a good answer, DeBow debated the question.

His words summed up the sentiments expressed by many of the Tuskegee Airmen in response to similar questions. "…his own mind was sufficiently clear to realize that he was flying for his country. He felt that however imperfect our democracy was, it's the only system that could open the way to perfection and he wanted no part of a Fascist future. He recalled thinking he was flying for his mother and dad, who struggled in menial labor to provide an opportunity for him. He felt he had a job to do for his country and his race, and just as Booker T. Washington and George Washington Carver proved themselves as educator and scientist, he might prove to someone that Negroes could become good pilots and officers."

Neal Loving with one of the airplanes of his own design. Loving was the first black and double amputee to be qualified as a racing pilot by the National Aeronautic Association and the Professional Racing Pilots Association. *Neal Loving*

Enoch Waters, staff member at the *Chicago Weekly Defender*, wrote, "Chicagoans discovered how deeply racial discrimination had permeated all aspects of aviation. Blacks were denied instruction, they experienced difficulty in renting aircraft and even those owning planes were often denied parking space and service at private airfields."

Many blacks were angered at refusals of airport officials to allow them to use established airport facilities. Following the lead of white pilots who used pastures and unimproved fields for flying, enterprising blacks found other ways to get to the sky.

"…In the Windy City in 1931, John C. Robinson, one of the earliest black pilots, and a small group of aviation enthusiasts (including Coffey and Willa Brown) formed the Challenger Air Pilots Association, one of the first black flying clubs in the United States. The Challenger group, barred from established airports in

Lieutenant General Benjamin O. Davis, Jr.
Class 42-C

No one can better personify Tuskegee's Heroes than Benjamin Davis, who has courageously lived a life that serves as inspiration for all. His message has not changed in fifty years. Aware of a supreme effort put forth by the Tuskegee Airmen, he urges the youth of today to follow suit. They, too, can achieve success, "...provided they really make an effort."

Benjamin O. Davis, Jr., entered the U.S. Military Academy in 1932. He was not the first black at West Point. Others, including Col. Charles A. "Follow Me" Young, graduated from the Army's prestigious academy prior to the turn of the 20th century. First in this century to attend and to reach flag rank in the Air Force, Lt. Gen. Benjamin Davis, Jr. followed in the footsteps of men like Col. Charles Young and those of his father, the Army's first black general officer, for whom he was named. From the start, the younger Davis showed remarkable strength of character as during his entire college career he endured the infamous

silent treatment. To his credit, he graduated in June 1936, 35th out of a class that numbered 276! Commissioned a 2nd Lieutenant, he married Agatha Scott Davis shortly thereafter.

Davis, a Commander of an Infantry Company, graduated in 1938 from Fort Benning, Georgia's Infantry School and transferred to Tuskegee Institute, Alabama, to become a professor of military science. A graduate of Tuskegee's first cadet class, he made further history as one of the first blacks to serve as a flying officer in the U.S. Army Air Corps.

In a later speech, he said, "In 1941, the Tuskegee Airmen benefited greatly from the political pressures applied by the black press, pressures that ultimately resulted in governmental actions which supported our long-held and burning desire to fly airplanes. True, even when the opportunity was presented to them, the Tuskegee Airmen had to grasp that opportunity and push it to its limits. This they did. ...I bow to the black press. …I also salute the several hundred Tuskegee Airmen who were trained and subsequently moved to Tuskegee Army Air Field to be the all-important foundation for all of the black combat units that were later to be activated."

Davis said, "This was not a simple case of training people to fly airplanes. The Air Corps called it an 'Experiment' because it involved not merely a technical problem, but a human relations problem."

As commander of the 99th Fighter Squadron, he and the unit were sent to North Africa in April 1943 and later to Sicily. Returned to the United States in October 1943, he assumed command of the 332nd Fighter Group, Selfridge Field, Michigan, with which he returned to the Mediterranean Theater two months later. "Our mission of escort," said General Davis, "was really the prime mission to carry out successfully and this we did. The 332nd became known as the best escort operator in the 15th Air Force. We never lost a bomber to enemy action of airplanes."

From June 1945 to the following March, he commanded the 477th Composite Group at Godman Field, Kentucky, later assuming command of Godman Field. In March 1946, he performed double duty as Commander of the 332nd Fighter Wing and Post Commander at Lockbourne Army Air Base, Ohio.

Integration of the armed services cannot solely be attributed to one individual, but Davis' contributions are significant. He and the men who responded so successfully to his leadership proved that racial integration was not only advisable, it was essential.

Historian Col. Alan Gropman wrote of the "watershed" of integration in the U.S. Air Force. He said, "He (Davis) told members of his unit that integration meant equality of opportunity not only for them, but for all blacks entering the Air Force. …He wanted his people to be proud because it was their performance that had accomplished this dream. He insisted they had nothing to fear from integration because they were fully qualified members of the Air Force, as accomplished and ready to perform the mission as any whites. …He had insisted throughout that his pilots train, fly and fight by the book and never deviate from the mission, and that opened a future worth having."

the area, first set up shop at Robbins Airport at the black township of Robbins, Illinois. There they cleared the land, built a hangar, and began to acquire aircraft."

A Registered Nurse and pilot, Janet W. Bragg made a generous contribution to the flying club by purchasing the first aircraft for use by club members. An educated woman like Brown, Bragg attended Spellman College and did postgraduate work at Cook County Hospital and at Chicago's Loyola University.

In 1933, she attended the Aeronautical University of Chicago where she studied Theory of Flight and other ground school courses. She completed flying instruction and flew from the small field in Robbins until the storm forced the move to Harlem Airport on Chicago's outskirts.

Bragg was instrumental in developing the first college preparatory Training Flying Program for blacks for the purpose of providing primary flight training for military pilot training. These programs were funded by the U.S. government to prepare candidates for military flying after graduating from college.

In 1942, Ms. Bragg attended Tuskegee Institute's Civilian Training School where, under Charles Alfred "Chief" Anderson, she prepared for a commercial pilot certificate which she received in 1943 at the Palwaukee Airport, Chicago.

Chauncey Spencer

Chauncey E. Spencer flew as a stunt pilot and parachutist, making a parachute jump at the March 1939 air show that Willa Brown helped to organize and publicize. The son of Lynchburg, Virginia's Anne Spencer, a brilliant, nationally-recognized poet whose works were included in the prestigious *Norton Anthology of Modern Poetry*, Chauncey was raised in a home that was a mecca to intellectuals Langston Hughes and Dr. George Washington Carver, artists Paul Robeson and Marion Anderson, and civil rights·activists Thurgood Marshall, Walter White, W.E.B. DuBois, Martin Luther King, Jr., and Mary McLeod Bethune.

An outstanding contributor in his own right, Chauncey Spencer, with Dale White, made a historic flight that brought national attention to the plight of black airmen in 1939.

Chicago's black aviators had organized the National Airmen's Association of America with Cornelius R. Coffey as president. …One objective of the NAAA was to act as an agency to keep the nation's active and

prospective black aviation enthusiasts informed on essential matters.

Another objective was to have historic national implications. In considering legislation setting up the CPTP, Congress had excluded black institutions. Furthermore, in planning its wartime emergency expansion, the Army Air Corps had made no provision whatever for incorporating blacks in its ranks.

This situation was noted by a black lobbyist for government employees, Edgar

G. Brown, who also kept tabs on legislative matters pertaining to America's black community. Brown alerted the NAAA and urged it to dispatch representatives to Washington to confer with officials on these vital matters. Toward this end, two NAAA pilots—Dale L. White and Chauncey E. Spencer—were tapped for a well-publicized flight from Chicago to Washington. The two arrived in Washington to learn that Brown had arranged a meeting with a little-known US Senator—Harry S. Truman. When Truman heard their objectives, he promised to help.

Enoch Waters wrote, "On May 9, 1939, two daring young black aviators flew a plane from Chicago to Washington, DC. That's not a remarkable achievement in these days of jet planes. …The fliers didn't set any records for altitude or speed. In fact, one could have beaten their time by train or auto. …They were daring because their aircraft was a single-engine, propeller driven, biplane ill-equipped for the trip."

Waters continued, "Three hours after takeoff they were grounded in a farmer's field in Sherwood, Ohio. Colleagues from Chicago rushed to their rescue and, after a day's delay, they were airborne again. They were officially reprimanded for landing their plane at night at Pittsburgh Municipal Airport without lights, navigational instruments and communication equipment. They had endangered other aircraft, they were told, and admonished as reckless fools.

"After (only) six and a half hours in the air, but two and a half days after their departure from Chicago, they arrived safely at Washington's National Airport where they were greeted by Edgar G. Brown. …To help the weary travelers fulfill their mission, there was probably no one in Washington better suited than the intense, goateed man they faced. Brown was a civil rights activist, indefatigable and persistent. Alone, he had desegregated eating facilities at the same airport through a sit-in and action in the federal courts. An untiring lobbyist on the Hill, he was known by and avoided by many Congressmen as he urged passage of legislation he felt was helpful to the black cause.

"…The mission of White and Spencer was to demonstrate that blacks were interested in and capable of learning to fly and to persuade the government to make certain that the CPTP was open to all citizens regardless of race. Brown set up meetings with persons he felt could be helpful. En route to one of these, the trio accidentally encountered a senator from Missouri —Harry S. Truman. The senator, like so many, didn't know there were any black pilots.

"…Their briefing impressed Truman to the point that he insisted upon going to the airport to see the plane which had been flown to the Capital. When he viewed the forlorn aircraft, made more so by contrast with other planes nearby, he congratulated White and Spencer. 'If you guys had guts enough to fly that thing from Chicago, I got guts enough to do all I can to help you,' he told them." *And* he did!

In his autobiography, *Who Is Chauncey Spencer?*, Spencer is outspoken in his support for a society that is color-blind. He has been quoted as having said, "Racism is on both sides of the fence."

Spencer now lives across the street from his boyhood home, recently declared a Virginia Historical Landmark, "The Anne Spencer Home." He was strongly influenced by his mother, but gives credit to his father as a role model.

He said, "He was what we called a quiet man. When he said something, it was worth listening. He had good, sound footage. He loved his family. He provided for his family. Even his son—who was a nuisance."

Spencer said, "You have to let people know what you're thinking. You can't let them control your thoughts. …You don't have to be a leader. Just keep on *being*."

Those who Came After: Loving, Jackson, Anderson, and Forsythe

Neal V. Loving

An engineer with an inventive mind, Neal V. Loving turned his fertile imagination and capacious intellect toward aviation at an early age. Born in Detroit in 1916, he was barely graduated from high school before having constructed his first full-sized flying machine—a glider that he designed and certificated with the CAA.

Loving soloed in a powered aircraft in 1939 and obtained his pilot certificate in 1940 as he simultaneously worked with the Detroit Recreation Department teaching young boys the fundamentals of aviation via modeling. He also taught an aircraft mechanics course at Aero Mechanics High School, Detroit City Airport.

Finding a weight and balance problem with his first glider, Loving designed and built a second—the S-1, N27775. Tested extensively, this glider went into production in Loving's small manufacturing facility, Wayne Aircraft Company—the first such production facility owned and operated by blacks; Loving and another pioneer airwoman, Earsly Taylor. Earsly later married Carl Barnett, Loving's

Lewis A. Jackson, the guiding hand in bringing Tuskegee primary flight training to number one ranking in the Southeast Army Air Corps. *Dr. Violet Jackson*

Herman
"Ace"
Lawson

Herman "Ace" Lawson, one of the first group of very welcome replacement pilots for the 99th Fighter Squadron—taking some of the heat from the initial cadre of flyers—arrived in June 1943 to augment the manning of the 99th. His was a circuitous route. Mistakenly, he and another member of the 99th were sent to Brazil. Although replacement pilots were desperately needed, it took a while to sort out the assignment and to rectify the orders. "Ace" finally joined the squadron in North Africa and completed a total of 133 missions overseas, flying P-40s, P-47s, and P-51s before he was returned to Tuskegee.

Lawson gained an enviable reputation as an athlete and as a photographer prior to joining the military service. He earned the nickname "Ace" as a football player at Marysville High School, California. He learned to build and fly real gliders, which aided in his eventual selection as military pilot, and he was one of the first blacks in northern California to receive a Private Pilot Certificate. After having served in the Civilian Conservation Corps for a year, "Ace" entered Fresno State at which he was an outstanding athlete—Collegiate Light Heavyweight Boxing Champion for two years, the first black to play four years of football at Fresno State, and a letterman in Basketball and Track. In addition, "Ace" was a professional photographer for the college.

As a First Lieutenant, "Ace" was Flight Leader, 99th Fighter Squadron, 332nd Fighter Group. He earned combat award honors for action over Greece, Germany, France, Czechoslovakia, Austria, Italy, Yugoslavia, and Romania. He said, "As Flight Leader, I never lost a pilot to the enemy while in action. I am very proud of that!" He returned to Tuskegee AAF in December 1944 after almost 18 months overseas.

Awarded the Distinguished Flying Cross, Air Medals, Commendation Medal, Unit Citation Medal, Award of Valor and Campaign Medals, "Ace" survived two crashes in P-40s, both due to engine failure, one on land and one in the Mediterranean Sea from which he was rescued by the U.S. Navy. His exemplary conduct and military awards tangibly proved that the racist slurs that cut viciously when he applied for Army Air Corps pilot training were totally unjustified.

"As the only black in his civil pilot training class at Fresno State College, Lawson went with some of his classmates to apply for the training. Upon reaching the interviewer, an Army Air Corps Major, the officer took one look at him and yelled, 'Get the hell out of here, boy! The Army isn't training any night fighters!'"

Artist La Grone said, "When 'Ace' learned that an Army Air Corps base was going to be built at Tuskegee, he volunteered immediately. He drove his car to the railroad station and left it there. He was so eager to be a pursuit pilot that he never gave a thought to more than a thousand dollars worth of photo equipment locked in his car's trunk. He had one thought—to fly! He lost it all. He never saw the car or the photographic equipment again!"

brother-in-law, moved to Jamaica and taught many Jamaicans to fly.

Little did Loving know, when construction reached 70 percent completion on his production model, the S-2, that he would be involved in a devastating crash. Demonstrating his glider for a group of Civil Air Patrol Cadets, he was tired and somewhat rushed. He spent the night repairing a damaged seat to the craft and, consequently, had very little sleep. Unaware that the airspeed indicator had stuck—unfortunately too *high*—circumstances began to be overwhelming. Turning from base leg to final approach, the glider stalled. Slamming into the ground just short of the runway, Loving's right foot was so badly damaged that it had to be amputated immediately. His left foot had to be amputated six months later.

What might have been catastrophic to another merely served to strengthen this man's determination. Loving says, "There is no success without enthusiasm." Despite his trauma, this courageous man was fitted with artificial limbs by February of 1945 and soon was able to walk without the use of crutches. By 1946, he was recovered sufficiently to start the Wayne School of Aeronautics and to serve as manager and flight instructor!

Loving designed a single-place, midget racing airplane known as the Loving WR-1, the *Love*. He flew it on the first test flight in August 1950. Painted cream and bronze and bearing Race Number 64, he entered the 1951 National Air Races, qualifying by demonstrating a 6-G pull-out and a 266 mile-per-hour dive. However, in another twist of fate, the propeller spinner came off during those qualification laps, cutting two inches from the tip of one propeller, causing a tremendous vibration, and ruining his chances for racing.

Undaunted, the innovative and intrepid pilot—once repairs were effected—won "Best of Show" at the 1954 Experimental Aircraft Association convention. He flew his airplane on an ambitious flight from Michigan to Jamaica in the West Indies, departing Detroit in December of 1953 and returning six months later. He and his low-winged, single cockpit craft flew through Atlanta, Fort Lauderdale, Havana, Cuba, and to Kingston, Jamaica, in 17 flying hours, covering 2,159 miles, 215 of which was over water. It was in Kingston that he met his wife, Clare, sister to his flight student Carl Barnett, to whom he was married in 1955.

Loving's *Love* is on permanent display in the Experimental Aircraft Association's Air Adventure Museum, Oshkosh, Wisconsin. Loving completed two more, the WR-2, a high wing tandem pusher with folding wings and WR-3, a low wing tandem roadable concept with folding wings.

In 1961, forty-five-year-old Neal Loving graduated from Wayne State University with a degree in aeronautical engineering. He obtained a position at Wright-Patterson Air Force Base, Dayton, Ohio, as an Aerospace Engineer and moved with his wife and family to the home in Yellow Springs, Ohio, in which they still reside.

Loving has recently published his autobiography through the Smithsonian Press, appropriately entitled, *Loving's Love: A Black American's Experience in Aviation.* More than memoirs—it is a compelling tale of heroism.

When asked to name the highlight of his life, Loving responded with humility, "It must be that at this time a prestigious publishing house has seen fit to publish my memoirs. I would never have dreamed that possible."

Many things in which this innovative man has involved himself have turned out to be not only possible, but also emulated by others. For example, SR-71 *Blackbird* pilot and Strategic Air Command Inspector General Pat Halloran, USAF (Ret), has flown his Loving's *Love* to Oshkosh, stopping at Springfield, Ohio, to show his craft to the inventive designer.

An inspiration, Neal Loving was named by the U.S. Air Force as the outstanding handicapped Federal Employee of the Year in 1968. The wording on the commemorative plaque read:

"It took great courage and desire for a Negro man to rise out of the slums of Detroit in the 1930s, only to be slapped down by an accident in which he lost both his lower limbs; come back to enter college and to be graduated; and then to engage in a new field of endeavor in which, through his own resourcefulness, he became an acknowledged national and international expert.

Mr. Loving is admired and respected by all with whom he comes in contact and is an inspiration to all who know him."

Dr. Lewis A. Jackson

When the world lost Dr. Lewis A. Jackson in January 1994, it lost a brilliant educator and pioneer aviator. Like Neal Loving, he started with airplane modeling and advanced to designing and flying his own creation; in Dr. Jackson's case, a hang glider. It was 1930 and he was 16, although his interest in airplanes dated from when he was barely three.

Starting powered aircraft flight training in 1930 and soloing in 1932, Dr. Jackson set the stage for his distinguished career that included becoming a commercial pilot, flight instructor, pilot examiner, aeronautical educator, aircraft designer, and university

Willie Howell Fuller

The battle to take the Italian island of Pantelleria was recorded in history as the first time that enemy territory was subdued by air power alone. In this, the second attempt at War to End All Wars, the air assault began 30 May 1943 and Pantelleria's surrender occurred within two weeks. Willie Fuller was one of those who helped to bring quick success to that campaign and others that followed. Before the end of World War II, Willie Fuller had flown 76 combat missions.

Born in Tarboro, North Carolina, Fuller was educated at Tarboro public schools and, in 1942, was granted a Bachelor's degree in Mechanical Industries at Tuskegee Institute. Taken into the cadet training program in 1942, he completed training as an original member of the 99th Fighter Squadron.

Fuller wrote, "I was a Gunnery and Advanced Student and a Pilot Instructor at Tuskegee Air Base."

Having completed the program in August 1942, he was commissioned a 2nd Lieutenant and sent overseas. He helped his 99th Fighter Squadron subdue Pantelleria, Salerno, and assisted in driving the German retreat. Some of his missions required bombing with fragmentation bombs, dive bombing, fighter bombing, flying escort for heavy and medium bombers, as well as strafing and fighter sweep missions.

Fuller served with U.S. forces until 1947. He was awarded the Air Medal with oak leaf cluster and promoted to Captain, U.S. Air Force Reserve, after serving overseas for fourteen months.

Married to Willie (Billie) Dunson, Fuller returned to Tuskegee Institute to complete a course of study that resulted in a degree in secondary education as a science major. In 1954, he attended the National Training School for Professional Scouting, Schiff Scout Reservation, Mendham, New Jersey. Assigned as a Professional Scout, West Georgia Council, he served the Boy Scouts of America (BSA), accepting a position with the South Florida Council, BSA, as District Executive serving youth in Dade, Broward, and Monroe counties.

A member of the Greater Florida Chapter Tuskegee Airmen, Inc., Willie Fuller devoted his life to the enrichment of the lives of youth.

Gene Carter, pilot and maintenance officer, served with the 99th Fighter Squadron in North Africa, Sicily, and Italy. His career included two tours with the Air Force ROTC Detachment at Tuskegee University. *Lt. Col. Gene Carter, USAF Retired*

administrator. Having purchased a Waco 10 aircraft, he earned his pilot's certificate, saving sufficient money to pay his way through college by barnstorming and crop dusting throughout his native Indiana and northwestern Ohio from 1932 through 1937, when he earned the Transport Pilot's Certificate.

In 1939, he received a bachelor's degree in education from Indiana Wesleyan as well as his Commercial certificate with Instructor Rating. He taught public school and, in 1940, went to Tuskegee Institute where, as the Army Air Corps 66th Flight Training Detachment came into existence, Dr. Jackson was appointed Director of Training. He added his Instrument, Multi-engine and Airframe, and Powerplant Mechanic's ratings and became a Certificated Ground Instructor.

According to *OX 5 NEWS*, February 1994, "As director, Dr. Jackson guided the school to high standards of performance and, on three different occasions, the students ranked first when compared to the other twenty-two schools in the Southeast Army Air Corps Training Command."

An article in the *Dayton Daily News*, February 1989, stated, "In 1940 military aviation was still young, but racial discrimination was old and white America didn't allow blacks to join the Army Air Corps. The first crack in that pattern of segregation came at Tuskegee Institute and at the nearby Tuskegee Army Air Field, Alabama. …(The Tuskegee Airmen's) achievements, and their demands for equal treatment as military professionals, are credited with causing the U.S. War Department to review its racial policies after the war. …Jackson was a civilian instructor at Tuskegee and became director of training for Tuskegee's Flight School, supervising 45 civilian pilots, several ground instructors and two dispatchers." Through his position, Dr. Jackson was directly instrumental in assisting with the breakdown of segregation in the military services.

With the goal of enabling the common man to fly, Dr. Jackson designed and built aircraft. His tenth,

Following twenty years of distinguished military service, Elwood "Woody" Driver excelled in his distinguished career with the National Transportation Safety Board (NTSB), the arm of the U.S. government that is concerned with safety and accident investigation for highways, railways, and airways. Woody served as Associate Administrator for Rule Making from 1967 to 1978 and accepted a presidential appointment to Vice Chairman of the NTSB, serving from 1978 to 1981.

In keeping with the excellence that typifies the Tuskegee Airmen, Woody Driver set standards for outstanding achievements. A native of Trenton, New Jersey, he graduated from New Jersey State College of Trenton in 1942 with a Bachelor of Science degree and from New York University, from which he received a Master's degree in Safety.

Woody graduated as a 2nd Lieutenant in October 1942 and served as a combat pilot and flight leader in the 99th Fighter Squadron. He completed 123 combat missions during which he was credited with one confirmed and one probable aerial kill.

He described his conquest: "February 1944, about ten plus FW-190s dove from 16,000 feet from an easterly direction and flattened out on the deck over Anzio. I was headed west at 6,000 feet. Before the FW-190 reached a position beneath me, I made a diving left turn and pulled out about 300 yards behind him and began firing. I continued to fire in long bursts, even though he was pulling away. My tracers straddled the cockpit and a sheet of flames burst from the right side. I last saw the plane burning and headed toward Rome at 50 feet above the ground. The firing was done from 500 feet down to the deck. I was slightly above at all times. During the time I was firing, a clipped wing Spitfire was also firing. He was to my right and ahead about 50 yards. I claim one Focke-Wulf 190 destroyed.'"

Driver was credited with aiding C.P. Bailey: In February 1944, Bailey's forward visibility was obscured by a layer of oil that covered his canopy. He was led back to the home field, Capodichino, by Driver, who remained on his wing and guided him.

After World War II, Driver served in Japan as Commander, Showa Air Base, and Director of Safety, Far East Air Logistics Forces. He was Personnel and Executive Officer, Chanute Air Force Base, Illinois. Woody was married to Shirley Martin of Dallas, Texas.

After retirement with the rank of Major and twenty years of active duty, Driver served for five years as Chief, System Safety Engineering, North American Aviation, California. Hired by the NTSB, his second career took flight. Honored by the NTSB, artist Roy La Grone was commissioned to paint this portrait of another true hero—Woody Driver.

La Grone said, "Woody was a dear friend. They had a big ceremony for his retirement and intended to unveil this painting during that ceremony. I was stuck in traffic and missed the unveiling, but walked in before the ceremony was over. It was a thrill for me to do it. He was quite a guy!"

Elwood
Thomas
"Woody"
Driver

Taught to fly by Chief Anderson and others in the CPTP, Tuskegee Institute, Mildred Hemmons (Carter), was the first woman to graduate from the program and acquire her pilot's certificate. Mildred is shown with Mrs. Eleanor Roosevelt. In the background in the Tuskegee "T" shirt, is Daniel "Chappie" James. With the First Lady and Miss Hemmons are G.L. Washington, left, and Tuskegee President, F.D. Patterson. Miss Hemmons applied to the Women's Airforce Service Pilots, but was denied because of her race. Once licensed, she had no aviation career opportunities available. *Lt. Col. Gene Carter, USAF Retired*

the J-10, was successor to his other experimental designs. With several unique characteristics, perhaps the foremost were light weight and its detachable wings. The craft was designed to be stored at home in a garage and towed to an airport for flight. In 1956 he created the Versatile I, developed with swing wings to serve as an airplane and a car and, in 1960, he created another experimental plane inspired by short-winged birds. The wings on this roadable plane folded.

One of the Tuskegee Airmen, Harold Sawyer, said of Lewis Jackson: "I thought he was one of the greatest stunt pilots I ever saw in my life. I was spellbound every time I saw him go up in a Waco to teach or demonstrate acrobatics. I always admired him. He was a wonderful instructor."

Phillip Lee, a Primary instructor at Tuskegee, said, "At Tuskegee, he would fly with the students periodically. He said, 'I will grade you instructors on what your *students* show me.' Lewis Jackson was small in stature, but a giant in knowledge."

Dr. Jackson was honored posthumously when Greene County Airport was renamed Greene County-Lewis A. Jackson Regional Airport, in Xenia, Ohio. Listed in *Who's Who* in America, in Ohio, in Aviation, and in Education, Dr. Jackson joins Brown, Bullard, Coleman, Coffey, Bragg, Spencer, Loving, Forsythe, and Anderson and the many other deserving pioneers of flight in the Black Wings exhibit of the National Air & Space Museum. It is a well-deserved tribute, but one that never would have been in the mind of the young Lewis Jackson as he took his first airplane ride in an OX 5 Swallow biplane at age fifteen.

Dr. Albert E. Forsythe and C. Alfred "Chief" Anderson

Dr. Albert Forsythe, born in Nassau, Bahamas, in 1897 and raised in Jamaica, B.W.I., attended Tuskegee Institute and graduated from dental school in Montreal, Canada. Like the other blacks denied aircraft rental opportunities, Dr. Forsythe managed to purchase an airplane, a Fairchild 24. He obtained flight instruction from Ernie Buehl and from Charles Alfred "Chief" Anderson, with whom he set history-making, record-setting flights.

In 1934, the pair departed Miami and flew more than 12,000 miles throughout the West Indies and South America. Given instructions on a landing in Nassau—the first ever for a land plane—they were told, "Fly to the lighthouse and make a right turn onto the road. Watch you don't hit the old windmill."

Elected to the Aviation Hall of Fame of New Jersey in 1985, Dr. Forsythe was also honored by the city of Newark, New Jersey. The Major proclaimed 7 February 1986 is "Albert Forsythe Day."

Dr. Forsythe found a lifetime friend in Charles Alfred "Chief" Anderson. It is no accident that Chief Anderson's name has repeatedly appeared in many of the preceding vignettes of the black pioneers of aviation—his influence has been vast and his accomplishments, along with an auspicious list of "firsts," have been many. Like the widening circles that emanate from a pebble tossed into a still pond, Chief Anderson has touched the lives of an untold number. He has truly earned the title of "The Father of Black Aviation."

Money was of little advantage to a black man in 1927. "Chief" Anderson had saved enough money to take a few flying lessons that year. He was twenty years old and as eager as any prospective pilot but no one wanted to teach a black anything about flying.

He said, "In the latter part of 1928-29, I had saved about $500. I borrowed an additional $2,500 from friends and bought a small plane. Then I had to depend upon any pilot who was kind enough to advise me and fly with me. …After being chased from various airports, I finally found a friend in Ernest Buehl, who served in the German Air Force in World War I and who migrated to this country and started an airport in Philadelphia known as the Flying Dutchman."

Chief managed to solo and to earn Private Pilot certificate No. 7638, in August 1929. In addition, he was the first black to receive a Transport license, in 1932; just one of his auspicious list of "firsts." Having met Dr. Albert Forsythe, Chief taught Forsythe to fly and they began a friendship that would last throughout their lives. Together they flew the first round-trip continental flight by blacks in 1934, then flew to Canada.

"I vaguely recall," said Anderson, " a write up in *TIME* about two 'blackbirds' who were attempting to fly across the United States. Forsythe and I were the first blacks to fly to the Bahamas and the West Indies and on to South America. We landed the first airplane *ever* in Nassau which, at that time, had no airport."

In the late thirties, Chief started a CPTP course at Howard University, Washington, DC, one of the six black colleges and the few other places at which blacks were accepted for flight training. He was soon lured to bring his instruction skills to Tuskegee Institute, site of the largest black CPTP enrollment, by then-President Dr. F.D. Patterson.

Chief was sent to Curtiss Reynolds Airport near Chicago, now the Glenview Naval Air Station, to complete a secondary training course for flight instructors. His classmates included black aviation pioneers Dr. Lewis Jackson and Cornelius R. Coffey.

Rufus Hunt, Air Traffic Controller, wrote in *The Chicago Defender*, "Graduation depended upon successful completion of the aerobatic course. Stearman PT-13s were used, with the student in the front seat. For some reason Anderson experienced difficulty in executing slow rolls, though the instructor in the rear seat performed them flawlessly. Anderson took the ship up, flying from the rear seat, and performed the maneuver properly as he knew he could. He discovered that the front controls had been rigged so that some maneuvers could not be performed properly. Someone purposely wanted him washed out of the course. This was a period when America did not afford equal opportunity to all her citizens."

Chief has devoted his lifetime to sharing his love of flight with as many young people as possible—black and white, male and female. He admitted, "I cannot say how many individuals I have taught to fly nor can I say how many hours I have spent in the air with this objective." Chief Anderson passed away in 1996.

Roughly 1,000 pilots who successfully completed the Army Air Corps Training Program, Tuskegee, Alabama, were under the tutelage of Chief Anderson in the primary phase of flight training. He was the man who taught America's first black fighter pilots how to fly.

But it was into a tightly woven tapestry of prejudice and discrimination that Chief Anderson flew the first airplane to Tuskegee Institute, Alabama, in 1939. With his accrued 3,500 flight hours, he was the first black pilot hired as an instructor. Like the other pioneer black aviators whose courage and determination could not be stifled, those with whom he shared a vision of equal opportunities in aviation, Chief Anderson brought to Tuskegee the ability to inspire his students with his passion for the air.

Challenged to create capable pilots while criticized by those who jeered and said that it couldn't be done, Chief faced the unknown with confidence. It was his task—and his honor—to be a leader in the Tuskegee Experiment.

According to Warren's Colorful Nicknames, Herbert Eugene "Gene" Carter, Chief of Maintenance and Combat Fighter Pilot of the 99th Fighter Squadron, was called "Pepie Le Moko" from Peter Lorie's character in the movie *Casablanca*. Small of stature but great in achievement, "Pepie" made outstanding contributions to the success of Tuskegee Airmen.

Carter kept a diary while on the front lines of World War II battle. His last few entries prior to returning to his beloved wife, Mildred Hemmons "Mike" Carter, spoke eloquently:

3-Mar-44: Dear Mike, I am sending you this diary which has eleven months of my daily doings overseas. Lane is leaving to return to Rose and America. You will find him a changed man. Please let that be an example to you on how this stuff really works on a brother. We are not the same. We can't be the same.

4-Mar-44: With Lane going, that leaves Ashley, Bolling, Carter, Custis, Hall, Fuller, Lawrence, Rogers, Wiley, Knighten, and Spanky. Only eleven of us of the 30 who left you eleven months ago. The going has been pure hell. There are many times when you have had enough of the sweat and blood, but the inner craving for flying and excitement makes you go again and again until finally your number comes up. And it will come. Yes, it will. Wiley, Hall, Fuller, Knighten, and I are past our required missions to return home, but medical finds nothing wrong, so we're still at it.

5-Mar-44: We are changing over to P-47s now. You should see me. I look like the man in the whale. I can nearly stand up and walk around in the cockpit, but I like the big heavy thing. Those 2,000 horses really sound good and I am just waiting to put those 8 fifties on a Jerry. We can carry a 1,000-pound bomb under each wing and a 500-pound or belly tank under the belly. A lot of stuff to drop on the blasted Jerries.

6-Mar-44: We lost Mike III yesterday. I guess I win. Three out of three and I am still here to talk of it. Jerry flak shot a whole section of the engine out. Was able to make a dead stick landing on the beach with the wheels down, so a new engine will fix her up again. But, by then, I hope to have named a P-47 'Big Mike.'

When asked, Carter admitted giving the nickname to his wife. He said, "She was a tomboy; a pilot. She graduated from the same CPT program that we did, although there was no opportunity for her to fly, once licensed. She wanted to be one of the boys, so she's 'Mike.'"

Born in 1919 in Mississippi, the fourth of ten children, Carter enrolled in high school at Tuskegee, living with his older brother and working toward entry to the Institute. A football quarterback, member of the track and basketball teams and choir, president of the Honor Society and his senior class, he proved his determination and excellence.

Graduating from the fourth class—12-F—that trained at TAAF, Carter was Cadet Captain of his class and, in addition to flying 77 combat missions, served as Maintenance Officer. He said, "The Tuskegee scene was set up because of outside political pressure, really to prove that we could NOT fly. Of course, much to their surprise, the young men who volunteered for the program were the cream of the crop. Ninety-five percent of them were college graduates or with three or four years in college—including the mechanics and all of the administrative and technical people. In response to those people who said, 'You Can't Do It,' we were damned determined that we were going to do it. And we did!"

After World War II, Carter went on to fly five different types of fighter aircraft including the F-106 fighter interceptor. His assignments included: Group Maintenance Officer, Lockbourne Air Force Base, Ohio; Flight Test Maintenance Officer, Wright-Patt Air Force Base, Ohio; Professor, Air Science, and Commander, Tuskegee Institute; Deputy Director, Military Advisory Group to German Air Force; Chief of Maintenance, Loring Air Force Base, Maine; and Professor, Aerospace Studies AFROTC, Tuskegee Institute.

Highly decorated, Lieutenant Colonel Carter earned his Bachelor's and a Master's degree in Education from Tuskegee Institute. After retirement from the Air Force, Lieutenant Colonel Carter served at Tuskegee as Associate Dean for Student Services, Admission and Recruiting. Continuing his counseling activities to this day, he personifies the "cream of the crop." He said, "We as a group of people must never let down our guard and think that this world is free and open. Every generation must forego some of the things we went through some 50 years ago. We must always keep America aware of the fact that you cannot judge a man or woman by the color of his or her skin. Judgment must be based solely on performance and capabilities as demonstrated from day to day."

Herbert
Eugene
"Gene"
Carter

41

CHAPTER TWO

The Tuskegee Experiment

It was an experiment bound for failure! Senior leaders of the U.S. Army Air Corps were simply unable to believe that blacks could learn to fly or to perform in combat; their training, their experience, and their ultra-conservative attitude prevented it. They had never heard of Eugene Bullard, or Bessie Coleman, or Cornelius Coffey, or Willa Brown, and they had never seen a black pilot in the U.S. Army Air Corps! They didn't read the black press, so they never read of the Banning-Allen or Anderson-Forsythe flights. What they believed was the prevailing attitude in the United States and, at least partially, the outgrowth of a 1925 Army War College "study" that argued vehemently that blacks were unfit to fly.

Roy La Grone said: "There were people in the War Department who actually believed that blacks couldn't fly. A paper written in 1925 was the 'Bible.' The treatment of Negro soldiers was based on that paper that said they were childlike, always fighting, couldn't be trusted—in spite of the history of World War I and all of the wars in which they had served!"

What that "study" didn't address was the courage, the capabilities, and the determination of those blacks who were to become Tuskegee Airmen.

1939

Clouds of war darkened ominously in Europe and in Asia. In the west, Hitler's speeches inflamed a German populace still reeling from the impact of the Versailles Treaty following World War I's surrender. Economic conditions were disastrous, yet his war-making capability grew in size and readiness. In the east, Japan's army expanded on the brutal conquest of

Manchuria and the Empire's navy built and trained for war in the Pacific. Around the world, stages were being set and early action developed.

In the United States, international policy debates focused on isolation from the rest of the world, the impact of the Neutrality Act signed into law in 1937. Politicians struggled with domestic problems created by the devastating Great Depression. Only a few political leaders were willing and able to fully recognize the import of the gathering war clouds and to support military preparations. United States military forces were small and ill-equipped for modern warfare demands; a major build-up in manpower and materiel was imperative.

Those few cognizant of the growing need for increased military and naval capability included some leaders of the black community and their spokesmen in the black press. They were painfully aware that, despite historical records that proved capabilities of black units and the heroism of individual blacks in previous wars in which the United States was involved, U.S. military leaders isolated them into separate organizations, relegated largely to menial tasks. Too many capable individuals were forced to serve in a Navy's Mess Branch, in the Army's road-building efforts, or on general fatigue duty. In World War I, only 11 percent of enrolled blacks were assigned to Army combat units. None were assigned to the Army Air Corps (AAC) except in menial jobs. Even the pilot who had distinguished himself while fighting against the enemy for the French had been refused entry into the U.S. Army Flying Service!

Black leaders also knew of obstacles faced by blacks trying to get into civil aviation and the resultant small numbers involved—in 1939,

---

top
It gets cold in a P-40 at altitude. Leather and sheepskin helped. Class 43B graduated William Griffin, Claude Govan, Walter Downs, James Polkinghorne, John Prowell, Roy Spencer, and William H. "Wolf" Walker. *Office of Air Force History, Maxwell AFB, AL*

bottom left
Benjamin O. Davis, Jr., first commander of the 99th Fighter Squadron, and Dr. Fred L. Patterson, President, Tuskegee Institute. *Office of Air Force History, Maxwell AFB, AL*

bottom right
Class 42F. This class was the first to go directly from CPTP into Basic Training: William A. Campbell, Willie Ashley, Langston Caldwell, Herbert Clark, George Bolling, Charles B. Hall, Graham Mitchell, Herbert "Gene" Carter, Louis Purnell, Graham Smith, Allen G. Lane, Spann Watson, Faythe McGinnis, James T. Wiley, and Erwin Lawrence. *Lt. Col. Gene Carter, USAF Retired*

Edward Creston Gleed

From Buffalo Soldier to Tuskegee Airman

In 1866, four black regiments were created, including the 8th and 9th Cavalry, "The Buffalo Soldiers." Two famous officers to have served in these regiments were Gen. John "Black Jack" Pershing and Gen. Benjamin O. Davis, Senior, the first black to become a general officer in the U.S. Army and father of the renowned leader of the Tuskegee Airmen. In 1941, Edward Gleed enlisted in the 9th Cavalry and was assigned to Military Intelligence. He went from "Buffalo Soldier" to Tuskegee Airman.

A graduate of Kansas University, Ed Gleed studied law at Howard University prior to entering Aviation Cadets in the U.S. Army. He graduated in December 1942 as an officer and pilot. It was an auspicious start to a military career that spanned three wars and offered thirty years of honorable service to his country.

In July 1944, escorting heavy bombers to Budapest, Gleed and other pilots of the 332nd encountered Bf-109s. Gleed was credited with destroying two. He assisted with the destruction of bridges, supply dumps, communication centers, oil refineries, aircraft on the ground and in the air. Credited with three "kills" and two "probables," Gleed was one of the fighter pilots with the 332nd Fighter Group who could have been declared an "Ace." He served as Squadron Commander of the 301st Fighter Squadron and as Operations Officer for the 332nd Fighter Group, and later for the 477th, in addition to his role in aerial combat.

Gleed held positions as Air Base Commander, Fighter Wing Operations Officer, and Deputy Wing Commander and Vice Commander, 30th Air Division (Defense-ADD), and served overseas tours in Japan and India. At retirement in 1970, he was serving as System Program Manager and Chief Administrator/Contract Negotiator of government computer programming contracts for two major Air Defense Operational Control Systems. After retirement from military life, he entered Southwestern University School of Law in Los Angeles and graduated in 1976 with Juris Doctorate degree. He was 60 years old.

Married to the former Lucille Graves, Colonel Gleed was the recipient of many awards and honor. He was decorated with the Legion of Merit, Distinguished Flying Cross, Air Medals with four Oak Leaf Clusters, Air Force and Joint Services Commendation Medals, and the French Croix de Guerre with Palms. Edward Gleed died, 25 January 1990, at the age of 73.

Prior to retirement as a Colonel, Gleed was a Command Pilot with over 6,000 flying hours. One of those hours was a unique flight—a solo in a P-39 Bell Airacobra—while a cadet at Tuskegee. Briefed on flight procedures and performance characteristics of the single-seat craft, but with no further instruction, Gleed climbed into the cockpit of the P-39, took off, and flew the aircraft for nearly an hour. He landed to the applause and admiration of his fellow cadets.

Gleed said, according to Charles Francis, "When we were in training at Tuskegee and in combat, we never gave it a thought that we were making history. All we wanted was to learn to fly as Army Air Corps pilots, fight for our country and survive."

Although the Honor Roll attests that not all of them survived, make history is exactly what Gleed and his fellow Tuskegee Airmen accomplished.

only 125 blacks held pilot's licenses, 94 of which were student certificates. They were also cognizant of the negative attitude toward blacks among the leadership of the U.S. Army and Navy. Blacks were totally excluded from military flying. Determined to ensure that black aviators were trained to fully participate in the battles that they believed were sure to come, these black leaders were to profoundly impact the future course of aviation and military history.

Two events with long-term implications occurred in 1939. The implementation of the Civil Aviation Authority's (CAA's) Civilian Pilot Training Program (CPTP), which was originally announced by President Roosevelt at the end of 1938, authorized the flight training of 20,000 college students per year. The program was to utilize existing facilities and to include a 72-hour ground school course and 35-50 hours of flight instruction, enough to qualify students for a private pilot's license. To be funded by the National Youth Administration, the CPTP was an early part of the national defense build-up undertaken under the leadership of the President as his concern for the deteriorating world situation grew.

As one result of the May 1939 Spencer-White flight from Chicago to Washington and the fortuitous meeting with Senator Truman, Congress authorized additional funds for the conduct of CPTP at several predominantly black colleges, for

Gradual expansion of the primary pilot training program at Tuskegee's Moton Field created a need for a new field for basic and advanced training. It was time to proceed with the building of a military airfield. *Office of Air Force History, Maxwell AFB, AL*

North American AT-6 Texan Advanced Trainers in formation over the Alabama countryside. Gunnery, instrument, and formation flying prepared advanced flight students for combat. *Roy E. La Grone*

the training of some black students at white colleges, and for training at two non-college flying schools, one of which was the Coffey School of Aeronautics.

West Virginia State, Delaware State, Hampton Institute, Howard University, North Carolina A&T, and, at the last minute, Tuskegee Institute were added to the list of colleges at which CPTP would be conducted. At the end of the first year, 91 percent of the black candidates, men and women, graduated, the same rate as that achieved by male and female white candidates. In spite of this record, the Army Air Corps still refused to accept black candidates for training.

The second event was the passage, in April 1939, of Public Law 18, an appropriations bill that included a provision for contracting with civilian schools of aviation for primary flight training. That provision was designed to provide relief for the AAC training facilities at Kelly and Randolph Fields in Texas that were threatened by rapid increases expected in numbers of trainees. Senator Harry Schwartz of Wyoming proposed an amendment providing for the lending of aviation equipment to one or more schools approved by the CAA for the training of black pilots. It was this portion of the passed bill that gave the AAC the most trouble.

Grayson Sandridge, Jesse Hawkins, and Herman White, Control Tower Operators, controlled the mass take-offs of the 332nd's fighters from Ramitelli Air Base, Italy. *USAF Photo, Roy E. La Grone*

Congratulations were in order for the first graduating class, George Spanky Roberts, Benjamin O. Davis Jr., Lemuel Custis, Charles DeBow, and Mac Ross. They provided leadership for the 99th Fighter Squadron, the first black flying unit in World War II. *U.S. Air Force Museum, Dayton, OH*

Largely ignoring the performance of black pilot candidates in CPTP, Air Corps leadership continued to cling to the false notions of the 1925 Army War College "study."

Public Law 18 posed a real dilemma. Did the Schwartz amendment really mandate the training of blacks as military aviators or did it merely require lending equipment to CAA-designated schools? For its part, the CAA designated North Suburban Flying School at Glenview, Illinois, for the training of black pilots; it wished to demonstrate "the adaptability of the Negro to flying instruction." The Air Corps continued to argue that because the contract primary flight instruction program was to be funded by the War Department and since the War Department maintained a strict policy of segregated units and since it made no provision for any black Air Corps units, there was really no need for facilities at which to train them. While the Glenview and Coffey schools of aviation instruction continued to participate in CPTP and, later, War Training Service (WTS) courses, neither was offered a contract for primary flight training by the Air Corps.

The legislative battle raged between the War Department and the AAC on one hand and the CAA, some members of Congress, and the national black leadership on the other. When the War Department finally

Roy La Grone and Freddie Hutchins were classmates at Tuskegee Institute. La Grone said, "Freddie and I worked together on the school paper. He was quite a character on the campus as he was as a pilot—a great morale builder. Not many realize it, but Freddie flew in three of America's wars."

A native of Blakely, Georgia, Hutchins graduated in Class 43-D. Each of his P-51 Mustangs were nicknamed "Little Freddie" I, II, and III respectively, a nickname that was often given to Hutchins himself. Although crashes put two of his assigned aircraft out of commission, "Captain Freddie" lived to tell the tales.

In July 1944, Hutchins joined other pilots to escort B-24s to a bombing raid over Vienna and, moments prior to reaching the Austrian city, Focke-Wulf 190s were sighted. The Red Tails climbed out of 28,000 feet to close in on their targets 8,000 feet above. The Jerries, in an evasion tactic, dove toward earth with Hutchins and the others in hot pursuit. Knocking four FW-190s out of the sky, Hutchins, Weldon Groves, William Green, Leonard Jackson, and Roger Romine took the honors.

In October 1944, the 332nd, as a component of the 15th Air Force, was assigned to strike enemy airfields in and around Athens, Greece. Germans were prepared for that low-level strike and put up a barrage of antiaircraft fire. As Hutchins pulled up off the target, he was blasted, "with a volley of flak. I looked to my right and saw that my right wing tip was torn off completely. Looking back, I noticed that my tail assembly was practically shot apart. Flak was hitting my ship and I scooted down into the seat to get the protection of the armored plates. But, a volley burst through the plane's floor into my leg."

Hutchins nursed his badly-damaged Mustang toward friendly territory and crashed into a small opening, knocked unconscious. He said, "When I came to, I was sitting at my controls with the engine of my plane lying several hundred feet away. My goggles were smashed on my forehead. My head was aching and my legs felt like they were broken."

Greek citizens pulled him from his craft and raised him onto a donkey. The agony was indescribable. The Greeks, trying to be helpful, took him to a doctor who rubbed him with homemade olive oil, bandaged him, and put him to bed. Driven nearly wild with flea bites, Hutchins vowed to leave the Greek village immediately. Although still in pain, he finagled a trip to the nearest city and a way to return to base. "Captain Freddie" was credited with having downed two enemy aircraft before he himself went down due to the enemy ground fire. He later received a Distinguished Flying Cross for his bravery. According to his fellow Tuskegee Airman, Chuck "A-Train" Dryden, "Freddie's aircraft was so badly damaged that it wouldn't climb to a safe bailout altitude and, as a result, he bellied in at over 250 miles an hour! The crash and subsequent run along the ground destroyed everything except the seat he was sitting on. When the wreckage came to a stop, 'Captain Freddie' came through.

Freddie F. Hutchins

Nurses Abbie Voorhies (Ross), Ruth Speight, Della Rainey (in cockpit), and Mencie Trotter during their flight orientation, a special part of their important duties at Tuskegee Army Air Field. *U.S. Air Force Museum*

decided to cooperate, it directed the Air Corps to confer with the CAA. Yet, when the Air Corps was finally ready to move forward, the War Department said that no action was to be taken with regard to the training of black pilots.

Thinking that the matter had been settled with the passage of Public Law 18, Senator Schwartz became frustrated with the delay. He made personal visits to General Arnold, the Air Corps chief, and to Brigadier General Yount, Chief of the Training Group of the Air Corps, to demand that black pilots be trained. Leaders of the black community, including Walter White, Secretary of the National Association for the Advancement of Colored People (NAACP); A. Philip Randolph, head of the Brotherhood of Sleeping Car Porters, the nation's largest black union; and Mary McLeod Bethune of the National Youth Administration, made similar visits, promising to continue to press the Congress for additional legislation if progress toward the training of black pilots for the military was not promptly accomplished.

In June 1939, at Congressional hearings on a supplementary military appropriations bill, Secretary of War Henry Woodring was severely challenged by Congressmen Powers of New Jersey, Ludlow of Indiana, and Dirksen of Illinois. They argued that something specific must be done to

establish facilities for training black pilots, proposing an amendment that would set aside $1 million of the 8 million requested for the expansion of military pilot training for the training of black pilots. They further proposed that this facility be set up at Tuskegee Institute in Alabama.

On 15 October 1939, while the issue of military training for black pilots continued in debate in Washington, CAA Chairman Robert Hinckley advised Tuskegee Institute President Frederick Patterson that Tuskegee had been approved for participation in CPTP. Flight training was to be conducted by Tuskegee and the Alabama Air Service, at Montgomery Municipal Airport, and required ground training was to be provided in cooperation with the Alabama Polytechnic Institute at Auburn. It is a credit to the leadership at Tuskegee, to the public works commissioner in Montgomery, as well as to the leadership at Auburn to have worked in such harmony during the late 1930s when segregation was so firmly entrenched in the Deep South. When training actually began in December 1939, problems became immediately apparent. Each student had to make a daily 80-mile round trip between Tuskegee and the airport and transportation was scarce. Feelings of triumph in finally starting a U.S. government sponsored program to train black pilots were mixed with frustration and exhaustion.

1940

Federal, state, and Tuskegee officials met in February 1940 to seek a solution. They agreed to lease a nearby plot of land. Tuskegee Institute donated $1000 and the students provided labor to build Airport No. 1 on the Union Springs highway. Although the field, which was referred to much later as Kennedy Field, was small and had limited facilities for three Piper Cubs used for training, it solved the critical problem of the commute to Montgomery. The good feelings associated with the move to Airport No. 1 were intensified when 100 percent of Tuskegee's first CPTP class passed the CAA written exams, many achieving nationally noteworthy scores. Their test results warranted CAA commendation and, in May, they completed their flight tests and received Private Pilot licenses. Their performance clearly demonstrated the serious dedication with which they approached their task as well as their ability to fly.

In recognition, in July 1940, Tuskegee was approved for a secondary course of flight instruction consisting of 240 hours of ground training and from 35 to 50 hours in the air. This second course, to provide additional training for certificated students from all of the black colleges, and to continue segregation of black students, would require use of a second airport. Once again, Alabama Polytechnic Institute at Auburn came to the rescue by allowing Tuskegee to use its small WPA-built field. On the 29th of July, Tuskegee's first instructor, C. Alfred "Chief" Anderson flew low over the Institute in a new 225-horsepower Waco, tail number NC 20970, painted in Army Air Corps' blue and yellow. Tuskegee's flying training program coordinator, Mr. G. L. Washington, said, "To us it sounded like a bomber."

With the full support of Tuskegee President Patterson, the administrative and negotiating skills of Coordinator Washington, and the flight training ability of Chief Anderson, the Tuskegee CPT program was off to a "flying start."

If the program's success was to continue, however, something else was needed—a new airport. In a brochure prepared by the Institute's Bureau of Public Relations in October 1940, Tuskegee's alumni were urged to contribute to the development of this airport. Titled "Wings Above Tuskegee," the brochure stated, in part:

"LANDING—THREE POINTS:

I. Coordinator Washington feels that Negroes have a *number one opportunity* to participate in the march of aviation, not only as civilian pilots, but as expert technicians and aviation mechanics.

II. President F. D. Patterson believes that with a modern, approved airport on its property, Tuskegee Institute can become an "Air Center" to provide training that will prepare Negro youth to qualify for this *number one opportunity*.

III. Tuskegee Alumni have decided that if Negro students wanted aviation enough to travel 80 miles a day to learn flying (and topped it off by making a national record), friends and Alumni should provide an airport.

When Tuskegee Alumni DECIDE, what happens? Look at the Alumni Bowl, at the concrete sidewalks, and, before long— at TUSKEGEE AIRPORT."

Long on enticing rhetoric, but short on the $200,000 needed, Tuskegee's Board of Trustees was concerned about raising the funds or about using Institute resources for something as potentially fraught with peril as an airport for primary training. They appealed to a number of potential sources— the final one The Julius Rosenwald Fund of Chicago.

In 1941, the Fund held its annual meeting at Tuskegee and Mrs. Eleanor Roosevelt, the First Lady and a member of the Rosenwald Board, was present. When told of Tuskegee's problems with aviation training, Mrs. Roosevelt took an immediate interest and asked to be taken for a flight so that she might have a better understanding. Over the strenuous objections of the Secret Service, Chief Anderson took her aloft in one of the Piper Cubs. She must have been favorably impressed, because soon after her flight, the Rosenwald Fund loaned the Institute $175,000 of the $200,000 required. Among those introduced to the First Lady during her visit was Miss Mildred Louise Hemmons, a Tuskegee student who was her college's first black woman to receive her private pilot's license through the CPT program. Mrs. Roosevelt maintained a strong interest in blacks in aviation from that day forward, a strong proponent for fair treatment of the Tuskegee Airmen and a favorite among them.

As the CPTP at Tuskegee succeeded, the fight with the War Department and the U.S. Army over the training of black military pilots dragged on. During a January hearing on another supplemental appropriations bill, Senator Bridges of New Hampshire again raised the issue of lack of training for black pilots. He cited the War Department's September 1939 reply to the application for military flight training submitted by Mr. Frank S. Reed of Chicago, a portion of which

Lieutenant General Benjamin O. Davis, Jr.

COL. BENJAMIN O. DAVIS, JR. CO. 332ND FG·99TH FS · 100 FS · 301 FS · 302 FS · ITALY 1944·1945

Lieutenant General Davis attended Air War College prior to being assigned Deputy Chief of Staff for Operations, Headquarters USAF, Washington, DC. In 1953, he completed advanced jet fighter gunnery school at Nellis Air Force Base, Nevada, and assumed new duties as commander of the 51st Fighter Interceptor Wing, Far East Air Forces (FEAF), Korea. Davis also served as Director of Operations and Training at FEAF Headquarters, Tokyo, then became Vice Commander, 13th Air Force, with additional duty as Commander, Air Task Force 13, Taipei, Formosa.

In April 1957 he arrived at Ramstein, Germany, as Chief of Staff, 12th Air Force, USAFE. With the transfer of the 12th Air Force to Texas, Davis served as Deputy Chief of Staff for Operations, Headquarters USAFE, Wiesbaden, Germany. Upon rotating to the United States he became Director of Manpower and Organization, DCS/Programs and Requirements, Headquarters USAF.

Assigned as Assistant Deputy Chief of Staff, Programs and Requirements, Headquarters USAF, Washington, DC, he remained in that position until becoming Chief of Staff for the United Nations Command and U.S. Forces in Korea.

In 1967, General Davis became Commander, 13th Air Force, Clark Air Base in the Republic of the Philippines and progressed to become Deputy Commander in Chief, U.S. Strike Command and Deputy Commander in Chief, Middle East, Southern Asia and Africa South of the Sahara, Headquarters, MacDill Air Force Base, Tampa, Florida.

Lieutenant General Davis was justifiably proud of the men that he led. He said, "...I give full marks to the Tuskegee Airmen themselves. They bore the brunt of it all. However, they could not have achieved their glorious record without lots of help. That help came from Gen. Noel Parrish, who held in his hand the key decision that blacks could fly airplanes at a superior level of proficiency; it came from "Chief" Alfred Anderson and his corps of Primary Flying School instructors, who performed their mission in exemplary fashion throughout all the

read: "It is regretted that the nonexistence of a colored Air Corps unit to which you could be assigned in the event of completion of flying training, precludes your training to become a military pilot at this time."

It was more "Catch-22"! There were no units, therefore no need for pilot training; there were no trained pilots, therefore no flying units could be created. Amendments, modifications, and additions to this bill and to the Selective Service Act of 1940 were developed and offered by Senator Bridges and by Senator Wagner of New York, by Congressman Fish

of New York and by Dr. Rayford Logan of Howard University, the Chairman of the Civilian Committee on Participation of Negroes in the National Defense Program, and by Mr. Charles Houston, a civil rights lawyer for the NAACP. They felt that they had made real progress when the Selective Service Act was passed by Congress. It contained a key provision, originally written by Dr. Logan, that there should be no discrimination, based on race or color, against any person in the selection and training of men. Unfortunately, however, this provision was offset by

years of TAAF's existence; it belonged to the hundreds of officers and airmen, who did the support job at Tuskegee Army Air Field, without which our combat effort could not have gone forward. I must also mention with admiration, the wives of the Tuskegee Airmen who suffered the privations of Alabama while they rendered the necessary support which their husbands needed and enjoyed through the long years of World War II."

A forward-thinking man in addition to being a great leader, Lt. Gen. Benjamin O. Davis, Jr., does not rest on the laurels of the successful Tuskegee Experience. He has forged an agenda for the present. He cautions about becoming excessively ambitious, but urges that we secure for all Americans the full benefits of American citizenship. He said, "We must not forget that this last point represents the reason, from its very beginnings to the present time, for the existence of the Tuskegee Airmen."

He continued, "...Today's youth has a greater capability within its own grasp to surmount difficulty than existed for the young people in the days immediately preceding World War II. Today's youth is blessed with the support of a body of laws that did not exist until President Lyndon Johnson achieved his long sought-for legislation that has meant so much to each one of us."

Lieutenant General Davis said, "The privileges of being an American belong to those brave enough to fight for them."

He would like to see the development of hope fostered in young people and a helping hand lent to the young. He urges that young people be helped to be positive in their attitudes toward their own individual progress and that they be shown that they have reason to believe strongly in the probability of success in their individual lives, "...provided they really make an effort."

Jack Rogers, 99th Fighter Squadron, flight planning for a cross country training mission. *U.S. Air Force Museum*

two others that gave the Secretaries of War and the Navy broad discretion in deciding who was to be considered fit for service and whether or not adequate facilities existed for their training.

Concerned that these provisions would be used to keep blacks from serving, several black leaders developed a memorandum of principles that addressed the integrated training and utilization of black officers, enlisted personnel, specialists such as doctors and dentists, women, and civilians to administer the Selective Service Act. This memorandum was presented at a late September White House conference that included President Roosevelt, Navy Secretary Frank Knox, Assistant Secretary of War Robert Patterson, NAACP Secretary Walter White, A. Philip Randolph, and T. Arnold Hill, assistant to Mary McLeod Bethune. Apparently, no one from the Administration mentioned that, earlier in the month, the President directed his War Department to prepare a statement that, "colored men will have equal opportunity with white men in all departments of the Army," or that the President raised the issue again in

a mid-September Cabinet meeting. As a result of the President's direction and the added impetus of the conference, a revised policy statement was approved on 9 October 1940. It was released to the press by a White House actively involved in Mr. Roosevelt's campaign for a third term as President.

While new policy maintained the "separate but equal" status quo, it did address a number of key points: (1) black personnel will be employed by the Army in proportion to their representation in the population at large; (2) black organizations will be established in all departments, both combatant and noncombatant; (3) training of black pilots, mechanics, and technical specialists will be accelerated; (4)

provisions will be made for the training of qualified blacks as officers; and (5) black reserve officers will be utilized on active duty. A major step forward, this revised policy, which did nothing to improve conditions in the Navy, was quickly followed by the nomination and appointment of the Army's first black general officer, the courageous and capable Benjamin O. Davis, Sr., commander of the Fifteenth Infantry in New York; the appointment of Maj. C. C. Johnson, an Army Reserve officer, as executive assistant to the Director of Selective Service to deal with racial matters; and the appointment of Judge William H. Hastie, Dean of the Howard University Law School, as civilian assistant

to the Secretary of War for Negro Affairs. Among Judge Hastie's first duties in this position was to refute a War Department Statement, issued in October, that blacks were being trained as pilots, mechanics, and technical specialists and that black aviation units would be formed as soon as there were sufficient trained personnel. He pointed out that the only training blacks were receiving was in CPTP at six black colleges and one civilian training school under the supervision of the CAA.

The final, and very positive, step taken in 1940 was the December submission of an Air Corps plan to establish a black pursuit squadron and several supporting elements. White officers and non-commissioned officers would form the initial cadre of supervisors and instructors; training of technical and administrative personnel would be accomplished at Chanute Field in southern Illinois and graduates of the CPTP courses would be selected for military flight training. A tumultuous year ended with a question: Where would the military flight training be conducted?

1941

When the Air Corps proposed a segregated training program for black pilots, Judge Hastie protested. He argued for the assignment of black primary cadets to a variety of contract schools, many outside the south, where white cadets were also

Charles Edward McGee

Ploesti was a target in August 1944 and 332nd Fighter Group pilots furnished escort for bombers of the 15th Air Force. In an attack planned for Austria, bomber crews and their Red Tails aimed at the destruction of an enemy airfield that impeded the advance of Russians toward Berlin. Seven Bf-109s appeared and a dogfight ensued. Kills were attributed to Luke Weathers and William Hill and, the following day, to John F. Briggs, William H. Thomas, and Charles Edward McGee.

Charles McGee, who is depicted in the painting

with his crew chief, Nathaniel "Nate" Wilson, spent his childhood in Ohio, Illinois, and Iowa. Sworn into the Enlisted Reserve in 1942, as World War II interrupted his education at the University of Illinois, he advanced to flight training at Tuskegee AAF. A member of Class 43-F, he graduated as a pilot and 2nd Lieutenant—the beginning of an auspicious military career that spanned three decades and three wars. Charles McGee served his country in World War II, Korea, and Vietnam.

McGee flew 136 combat missions with the 332nd Fighter Group, completing his combat tour and returning to the United States on 1 December 1944. After being trained in twin-engine powered aircraft and assigned as a multi-engine instructor, a position he held until Tuskegee Air Field closed in 1946, McGee elected to remain in the service at the end of World War II. He reported to Lockbourne Army Air Base, Columbus, Ohio, as Base Operations and Training Officer.

In 1948, he satisfactorily completed the Aircraft Maintenance technical course, Chanute Field, Illinois, and accepted assignments to the 301st Air Refueling Wing, Smoky Hill Air Force Base, Kansas; to March Field, California; and to Clark Field, Philippine Islands. As a member of the 67th Fighter-Bomber Squadron, Colonel McGee flew P-51 aircraft during the Korean War. Credited with 100 missions, he was awarded the DFC, the Air Medal, and promoted to Major.

Returning to Clark Field, McGee took command of the 44th Fighter-Bomber Squadron and transitioned into jets—the Lockheed P-80 Shooting Star. After completion, in 1953, of Air Force Command and Staff College, Maxwell Air Force Base, Alabama, McGee served in the Air Defense Command, flew the F-89 Scorpion interceptor jet and was promoted to Lieutenant Colonel.

Offered a regular commission, McGee transferred to Italy in 1959 to command the Luigi Balogna Sea Plane Base and the 7230th Support Squadron in conjunction with deployment of the Jupiter Missile, NATO. He returned to the United States in 1962 and served at Minot Air Force Base, North Dakota, and at Richards-Gebaur Air Force Base, Missouri.

Tapped to serve in his third war in 1967, McGee completed Tactical Reconnaissance and training and was appointed to command the 16th Tactical Air Command Reconnaissance Squadron, Tan Son Nhut Air Base. He flew 172 missions during the Vietnam War and was awarded, with other air medals, the Legion of Merit.

From Southeast Asia, McGee was assigned as a Liaison Officer serving USAEUR and the 7th Army in Germany. He was promoted to Colonel and completed his European tour with the 50th TAC Fighter Wing as Chief of Maintenance. He was assigned as Director of Maintenance Engineering, Air Force Communications Service and Commander of the 184th Air Base Wing, Richards-Gebaur Air Force Base, Missouri, in 1971. He retired 31 January 1973, a Command Pilot with over 6,300 flying hours.

Colonel McGee's awards include: Legion of Merit with OLC, Distinguished Flying Cross with 2 OLCs, Bronze Star, Air Medal with 25 OLCs, Army Commendation Medal, Air Force Commendation Medal with OLC, Presidential Unit Citation, Korean Presidential Unit Citation, the Hellenic Republic World War II Commemorative Medal, and several campaign and service ribbons.

Graduation—Mildred Hemmons (Carter) pins new rank on 2nd Lt. Herbert Eugene "Gene" Carter. Lieutenant Carter was a key player in the 99th Fighter Squadron's combat success.
Lt. Col. Gene Carter, USAF Retired

being trained. He also suggested that Tuskegee Institute, which had already established a flying training program and which had become active in locating necessary land, was interested in cooperating with the officials at Maxwell Field in the development of a military flight training facility at Tuskegee. Noting that Kelly, Randolph, Moffet, and Maxwell Fields were already feeling the pressures of rapidly growing student populations, the Air Corps argued that establishment of a training facility at Tuskegee would result in the least delay in starting a military flying training program for blacks. In January 1941, although he continued to oppose the "separate but equal" approach, Judge Hastie reluctantly withdrew his formal objections to the plan.

War Department, Army and Air Corps leaders continued to privately express reservations about the ultimate efficacy of black aviation units and the Air Corps continued to refuse flight training applications from black candidates until Yancey Williams, a Howard University student backed by the NAACP, filed suit to compel the Army to accept him. Finally, in March, the Air Corps began accepting applications and Selective Service headquarters

On the flight line at Tuskegee Army Air Field: Wilson Eagleson, Harold Sawyer, Heber Houston, James Brothers (later killed in a Stearman PT-17 Kaydet), and Arnold Cisco.
Harold Sawyer

the appropriation of just over $1 million to begin construction of Tuskegee Army Air Field.

A plan for military flight training at Tuskegee was consistent with that established in 1939 under Public Law 18. It was the approach used throughout the Air Corps during World War II and for many years afterward and the approach whereby 191,654 trainees were awarded their pilot wings between January of 1941 and August of 1945.

The flying cadet program was a demanding one everywhere and no less so at Tuskegee where, although the program was generally consistent with what was done elsewhere, there were major differences. Following acceptance into the program, white cadets were sent to one of three classification and pre-flight centers—Nashville, Tennessee; San Antonio, Texas; and Santa Ana, California. Black cadets were sent to Tuskegee. Physical exams and psychomotor tests were given and white cadets were classified for pilot, navigator, or bombardier training depending upon their aptitudes. Black cadets were limited to training as pursuit pilots. Both groups received their initial indoctrination to military life— "G.I." haircuts, shots, and introductions to marching, guard duty, and "K.P." They then entered nine weeks of pre-flight ground school, for the great majority their first introduction to airplanes and flight concepts. The cadets also were introduced to another kind of pressure that would challenge their stamina throughout the program.

Lieutenant Colonel Herbert Eugene "Gene" Carter, USAF (Ret), described, "The hazing was terrific. They put you on a brace and kept you standing against a wall. Or they put you on infinity. Your beds and your room never passed inspection. You always had to walk tours on Saturday. As a matter of fact, between January and June, when I received my commission, I had one open post. All those other weekends I was walking tours for demerits."

Assigned to their primary flying schools—civilian contract flight schools, cadets received approximately sixty hours of flying training in nine weeks, in Stearman PT-13s, Ryan PT-16s, and PT-20s or Fairchild PT-19s. Instructors were civilians, most of whom had completed specialized training in Air Corps training methods. The training aircraft were owned by the Army Air Corps and civilian operations were monitored and supervised by detachments of Air Corps personnel. At Tuskegee, the 66th Army Air Corps Training Detachment, initially commanded by then-Captain Noel F. Parrish, performed this role for the civilian instructor corps led by Chief Anderson and Lewis Jackson.

announced that pilots would be selected from among those who had completed the secondary course offered by the CAA. Soon thereafter, the War Department wrote to Tuskegee's Dr. Patterson to advise him of the plan to proceed immediately with the establishment of the 99th Pursuit Squadron to be made up of 400 enlisted men and officers, 33 pilots, and 27 planes, and to further advise him of

After successful completion of fundamentals of flight in primary, cadets moved on to basic flying schools which made military pilots of them. Flying in North American BT-9s or BT-14s, or in Consolidated Vultee BT-13s, all aircraft of greater weight, horsepower, and speed, cadets learned to navigate, to fly on instruments, to fly at night, and to fly in formation. They received seventy hours of flight training in approximately nine weeks from military instructors, who were reported to have remarkably little patience, and they learned to deal with two-way radio communications and a two-pitch propeller.

As each cadet successfully approached completion of basic, instructors decided whether he would go to single-engine or twin-engine advanced training. Since there were no twin-engine advanced trainers at Tuskegee, all cadets who completed basic flight training satisfactorily went on to fly single-engine fighters, whether or not that was what they were best suited for. This was a key difference in the training afforded to black cadets; another was that they remained at Tuskegee for advanced training instead of going to another base with a "clean slate."

Cadets who went on to single-engine training flew North American AT-6s for an additional seventy hours, concentrating on aerial gunnery and combat maneuvers while they increased their skills in navigation, formation, and instrument flying. Twin-engine advanced training was conducted in Curtiss AT-9, Beechcraft AT-10, and Cessna AT-17 aircraft. Here the concentration was on instrument flying, on flying at night and in formation, as well as on learning to fly an aircraft with more than one engine. Cadets who survived the three very demanding phases of flying training—literally as well as figuratively, since the loss rate due to accidents and to academic or physical problems with flying was about 40 percent—were awarded pilot wings and appointed to initial ranks of either 2nd Lieutenant or Flight Officer.

White cadets who "washed out" due to academic or flying deficiencies were offered opportunities to train for another flying specialty or to serve in other career fields. Black cadets who were eliminated before the war began were simply sent home. After Pearl Harbor, black eliminees from flight training were appointed Privates and retained. It wasn't until 1943 that blacks who "washed out" were given a chance to train as navigators or bombardiers.

White cadets who earned their wings entered into transition training for the type aircraft they would fly in combat. Some went to squadrons scheduled for overseas deployment and others to replacement training units for subsequent assignment to units already in combat. Their transition training took approximately two months. In stark comparison, early graduates of the military program at Tuskegee stayed in transition training at Tuskegee for almost a year. Their frustration mounted even as they sharpened their skills. They wanted to get into the fight! They had to fight for the *right* to fight!

A final, and critical, difference between the flying training available to white cadet candidates and that for blacks was in the numbers of positions available. Quotas for cadet classes at Tuskegee were set initially at fifteen and lowered, for some time, to ten. Numbers of white cadets were limited only by the available spaces in training; new bases to accommodate ever larger numbers were continuously under construction across the country.

Less than 1/2 of 1 percent of those awarded pilot wings during World War II were black. Many (like artist Roy La Grone!) who successfully completed CPTP were drafted to serve in non-flying roles before their applications for flight training were acted on; the AAC was unable or unwilling to provide additional training facilities. Segregation in flying training clearly denied the United States the talents and skills of many highly qualified black aviators. Air Corps leadership continued to believe that the Tuskegee "Experiment" would fail and they did all they could, despite the watchful eyes and political pressures of the Administration, Congress, leaders of the black community, and the black press, to make sure that it did!

Cadets who came to Tuskegee in 1941, and support personnel who worked so tirelessly to put CPTP and military flying programs in place, were even more determined to make sure that the "Experiment" succeeded! As one of them expressed it, "We were determined that this experiment was going to be a failure of THEIR endeavor, not OURS! We were going to do our best. And we did. We studied at night, we pulled together and we did everything that was expected of us and a little bit more."

Contractors and construction workers began building Moton Field, Airport No. 2, about four miles from Airport No. 1. On 23 July, another group, headed by the black-led firm of McKissack and McKissack, went to work on Airport No. 3, Tuskegee Army Air Field (TAAF). All primary training of black pilot candidates would be conducted at Moton and basic and advanced training would take place at TAAF.

*Harold
Emmett
Sawyer*

Harold Sawyer's parents were fortunate. Their three sons graduated from college and served overseas in U.S. military service during World War II. All returned to the family business in Columbus, Ohio, where Harold lives today with his wife, the former Gertrude Adams. Harold, a pilot with the 332nd Fighter Group, was at Tuskegee Institute when CPTP was offered and graduated in May 1942. Harold said, "I took advantage of that opportunity, got my private pilot license, and was working on my commercial when war broke out. I had to register for the draft, but hoped to get called for the Air Corps first!" Sawyer went to St. Louis for a course in aircraft riveting. When told to report to TAAF for flight training, he said, "I was tickled to death."

Acceptance for aviation cadets was based, in part, on physical examinations at Maxwell Field. Sawyer was underweight and had been examined twice. He said, "The doctor knew I wouldn't have many more chances and said, 'Next time you are scheduled, get a sack full of bananas. Eat them on your way here and drink all the water you can.' That did the trick, but it took years after that before I could ever face a banana again."

On 20 April 1943, graduating in Class 43-D—with C. P. Bailey, Wilson Eagleson, V.V. Haywood, Freddie Hutchins, Luke Weathers, C. I. Williams and others—Sawyer became a 2nd Lieutenant and pilot. He said, "We started out with fifty, the largest class to start into training," he said, "but only nineteen of us graduated. They had a quota system. They were NOT going to let but so many graduate. Period. It was frustrating, because some of the guys went clear up to the night before graduation and then washed out. You were on a tightrope! As you went onto the flight line for your hour, you never knew whether you were going to be coming back the next day or not. I never heard of anybody getting washed out of ground school, but they whacked you on that flying. We know there were just as many good pilots that washed out as made it through."

Assigned to the 301st Fighter Squadron, Sawyer flew P-39s, performing coastal and harbor patrol. He said, "When we were transferred to the 15th AF, that's when we were flying long range missions, bomber escort in P-51 Mustangs." He respected the awesome responsibility shouldered by their commander, then-Colonel Benjamin O. Davis, Jr. "It wasn't easy to keep a tight rein on a young bunch of fighter pilots! When we first got to Ramitelli, Lee Rayford, Colonel Davis, and I were flying a familiarization flight in the P-47. I had never been one to eat breakfast , but had downed powdered eggs that day. When we climbed above 13,000 feet, I got so sick, I threw up! I asked, 'Can we drop down? I gotta get this mask off.' Colonel Davis came right back, immediately, with, 'What would you do if you were in combat?' He was right. I stayed up there until we landed!"

Sawyer discovered a lot about aerial combat. A quiet and gentle man, he doesn't flaunt his 130 missions, that he was awarded the Distinguished Flying Cross and Air Medal with seven Oak Leaf Clusters, or that he shot down a FW-190 and a Bf-109. The DFC was awarded for a bomber escort flight over Linz, Austria, in which he was credited with the Messerschmitt and with damaging two others. A newspaper report stated, "Captain Sawyer has demonstrated this same extraordinary courage and devotion to duty throughout his outstanding record of seventy-plus successfully completed missions against the enemy." Honored by his hometown of Columbus, then-Mayor Rhodes proclaimed Monday, 22 January 1945 as "Captain Harold Sawyer Day."

Sawyer felt that the personnel of the 332nd Fighter Group were among "the best-kept secrets of World War II. If you talk to some of our guys who were prisoners of war, they say that German interrogators knew more about our Group than Americans did."

It is high time that he and his fellow heroes are appreciated. Their mastery of fighter combat is cause for celebration and includes some of their most courageous and memorable hours.

As engineers started to level the hilly, wooded site, the AAC assigned Major James A. Ellison to command the new base. He made considerable progress in getting the initial organizational structure into place but he proved to be a serious challenge to the area's white officials and he was replaced, less than six months later, by Col. Frederick Kimble. A West Point graduate with more than 24 years' flying experience, Colonel Kimble was also a senior Army officer with experience and attitudes that helped him to be more "politically correct" in dealing with local officials than his predecessor. He was a strong and unrelenting segregationist!

Tuskegee Airman Col. Lee Archer from New York City tells of his arrival: "The town of Tuskegee was probably one of the strictest segregationist towns. I arrived on Christmas Day, 1942, at the small station called Chehaw where I was supposed to be met by a contract bus company which didn't show up. So, I had to walk with my bags to the base—quite a distance. I find on the base that it is a segregated base. Places for whites, places for coloreds, white fountains, colored fountains. All this was a complete shock to me."

In October, the first black enlisted members of the 99th Pursuit Squadron arrived. Most had come into the Army Air Corps from predominantly black college campuses and many were qualified to be pilots and officers. Some went to Fort Monmouth, N.J., for communications training and some to Lowry Field in Colorado to be trained in armament. Many had been sent to Chanute Field for training as aircraft mechanics and in administration. An initial cadre of non-commissioned officers was transferred there from the all-black 24th Infantry Regiment to provide military supervision and training for the new recruits whose technical training was to last from nine months to a year. Many of those recruits had backgrounds in engineering and other technical disciplines. All were highly motivated and most were more qualified to teach the established courses than their white instructors.

To the amazement of the school's senior administration, the students excelled in their technical and military training; they often overwhelmed the school's ability to provide challenging work and, as a consequence, often had little productive to do. Because their military bearing and precision marching was so good, a drill team was formed. Regular performances made the team the talk of the base and of the town, and the base commander wanted to keep them assigned even after their training was officially complete. He lost out in his bid to retain them, however, and the entire group

Artist Roy La Grone's portrayals of black heroes in aviation are echoed in the terminal building, Lambert-St. Louis International Airport. A mural, entitled Black Americans in Flight, is "a Tribute to African-American Achievements in Aviation from 1917 to the Present. Painted by Spencer Taylor and assisted by Solomon Thurman, the mural was commissioned in 1986 and dedicated August 13, 1990." It is a pictorial story of American history.

Wendell Pruitt, a St. Louis resident and prominently featured in the mural, brought honor to his home town; 12 December 1944 was declared "Captain Wendell O. Pruitt Day." Called a "Tuskegee Pilot of Distinction," it captions, "Lee Archer and Wendell Pruitt, considered the best one-two punch of the 332nd Fighter Group." Pruitt is highly praised for an unheard-of event—he and Gwynne W. Pierson, 25 June 1944, together sank a German destroyer with only their aircraft machine guns, the feat shown in La Grone's painting.

A graduate of the Class 42-K, Pruitt was more than a natural pilot, he was a daring pilot. With 70 combat missions, he was credited with having downed three Bf-109s.

Held in such high regard as a pilot, Pruitt's death was hard to accept. Colonel Gene Carter said, "Pruitt was killed after the war out at Kennedy Field on Union Springs Highway. Like most of the returnees, he was one of the flight instructors down on the flight line. It was a Sunday and he was flying a T-6 with an enlisted man in the back seat. Pruitt came across the field in a low pass and pulled up into a chandelle. He went around and made another low pass, starting a slow roll as he pulled up. When he got into the inverted position, the T-6 just dove into the ground.

"They theorized that the enlisted man either fell forward or grabbed the stick and Pruitt couldn't correct. They even wondered, although they weren't able to tell, whether a seat belt was fastened around the passenger. The aircraft completely burned. Part of what they found was the unsecured buckle of the seat belt. No one will ever know what really happened."

Wendell Pruitt—One of Tuskegee's Finest. He lived to fly. He died flying.

Wendell O. Pruitt

Sports played a major role in keeping the airmen fit and close to the base. Here, a solid line drive in an inter-unit baseball game at Tuskegee Army Air Field. *Office of Air Force History, Maxwell, AFB, AL*

was transferred to Maxwell Field to await assignment. The AAC created another "Catch-22"! Since, in the "separate but equal" army, newly trained men could only be assigned to the single segregated unit—the 99th Pursuit Squadron—and since the base for that unit had not reached a stage in its construction where it could accommodate and utilize their skills, they had to wait at Maxwell until it could. Again, with little or nothing to do, it is a tribute to their character and leadership that trouble didn't break out.

Highlights of 1941, however, were on 19 July when a ceremony marking the official opening of military pilot training to black cadets was conducted at the monument honoring Booker T. Washington, the focal point of the Tuskegee Institute campus, and on 25 August, when the first class had its first military flying lesson.

That first class, 42-C, carefully selected and highly competitive, consisted of twelve cadets and one officer. The officer, Capt. Benjamin O. Davis, Jr., was the first black to graduate from West Point in the 20th century and the son of Brig. Gen. Benjamin O. Davis, Sr., the U.S. Army's first black general officer. He came to Tuskegee from Fort Riley, Kansas, and a series of Infantry assignments following his graduation from USMA at West Point in June 1936. Fully qualified mentally and physically, he had tried repeatedly for an Air Corps assignment. He received one written refusal

from the Army Chief of Staff in which it was stated that there weren't any Negroes in the Air Corps and there weren't going to *be* any.

In July 1941, Davis' dream of flying for his country was finally starting to come true. His cadet classmates, all of whom had also been forced to fight for the right to fight, were also college graduates; one, Theodore Brown from New York City, was only one course short of a doctoral degree. They came from across the country and they brought with them the typical mix of hopes and plans. Seven were eliminated for flying deficiency. One was thought to have eliminated himself when he learned enough about flying to conduct an aerial spraying service for farmers in his native Texas. Since war had not been declared, there was no obligation to continue in the Army. Those who washed out went home.

Graduating with Captain Davis were: Lemuel Custis, a former police officer from Hartford, Connecticut, and a Howard University alumnus; Charles DeBow, a Hampton Institute graduate from Indianapolis, Indiana; Mac Ross of Dayton, Ohio, a West Virginia State College graduate who had worked as an inspector at an iron works in his home town; and George "Spanky" Roberts, a native of Fairmont, West Virginia, and an alumnus of West Virginia State College. DeBow, Ross, and Roberts had completed CPTP. Like all fledgling pilots, they worked hard and put in long days. Some, like Spanky Roberts, seemed to have natural ability to control an airplane throughout a range of maneuvers. Others, like Mac Ross, could fly as well as Spanky, but had trouble with the course's book work portion. Each sweated in the humid Alabama heat and keenly felt the tension and anxiety of the demanding program. When Captain Davis soloed on 2 September 1941, he became the first black officer to officially fly alone in an Army Air Corps aircraft. In that segregated world, this was a major milestone!

At Tuskegee, the first class received their primary training under Charles Alfred Anderson, the man who became known simply and affectionately as "Chief." Born in Bridgeport, Pennsylvania, Chief was a true aviation pioneer. His commercial pilot's license, earned in 1932, was an unheard of accomplishment for a black man, and, with Dr. Albert Forsythe, he established new records in cross-country flying. An extraordinarily capable flight instructor, he came to Tuskegee in July 1940 from aerobatic training at Coffey's school in Chicago to initiate the secondary CPTP course.

Selected to be the chief civilian flight instructor at Moton Field when it opened on 23 August 1941, Chief had accumulated over 3,500 flight hours, much of it instructing new flight students, when he began at Moton. He was still flying, often with one of his former students, over 50 years later! To describe Chief as beloved of the Tuskegee Airmen is appropriate; they speak of him with the greatest respect and admiration.

Working with Chief at Moton Field was a capable group of black civilian instructors, many of whom had completed the advanced CPTP course and the AAC course for instructors. They included, to name only a few: Milton Crenshaw, Roscoe Draper, Charles Foreman, Charles Foxx, Lewis Jackson, Daniel "Chappie" James, Dennis Lee, Wendell Lipscomb, Adolph Moret, Claude Platt, Sherman Rose, and Perry Young. The aircraft mechanics and support staff at Moton Field were all civilians; among them was Miss Marjorie Cheatham, a young black woman who was a certified aviation mechanic, and Pappy White, who was Tuskegee's senior mechanic. White described his coming to Tuskegee this way: "I came here in 1941 to finish up on my degree. I had attended here earlier. I was changing my major and visited the airport with Percy Sutton at the old field. I observed Percy's work and ended up never going to school, but becoming chief mechanic and obtaining my A&E license in the latter part of '41 and '42." Many went on to outstanding aviation careers in both civilian and military roles. Lewis Jackson became a college president, Chappie James was the first black four-star General in the USAF, and Percy Sutton served as the Borough President of Manhattan in New York City.

Also working with Chief Anderson was Capt. Noel F. Parrish, Commander of the 66th Army Air Corps Training Detachment. The Detachment supervised the primary flight training school for the Air Corps and Captain Parrish made notable contributions to its success before he moved on to Tuskegee Army Air Field, first as Director of Training and, on 26 December 1942, as Commander, succeeding Colonel Kimble. Captain Parrish was born in Lexington, Kentucky, on 11 November 1909. He graduated from Rice Institute in 1928 and enlisted in the Army as a Private in July of 1930, serving for a year with the 11th Cavalry. When his application for flight training was approved, he entered primarily as a cadet at March Field, California. He was a member of the first basic flying school class at newly completed Randolph Field in San Antonio, Texas, and was awarded pilot wings following completion of advanced training at Kelly Field in July 1932, the same year Chief got his commercial "ticket."

Initially assigned to the 13th Attack Squadron at Fort Crockett, Texas, Parrish was transferred to the First Provisional Transport Squadron, Wright Field, Ohio, in February 1934. Commissioned a 2nd Lieutenant in the Regular Army of the United States in July 1935, he rejoined the 13th Attack Squadron. Following a subsequent tour as an instructor at Randolph, Parrish was sent, in the summer of 1939, to assist in the establishment of the primary flying school at Glenview, Illinois, one of the original civilian contract schools established under Public Law 18. He was re-assigned to Tuskegee as Commander of the 66th in May 1941 and moved to Tuskegee AAF with Class 42-C in November.

Captain, later Brigadier General, Parrish was unique among his peers; his soldierly approach to his duties, his successful efforts to understand and to relate to the perspective of his black command, the leaders of the black community, and the people of Tuskegee enabled him to achieve his objective. He wanted nothing more than to train individuals and an organization fully capable of fulfilling the mission assigned.

Later, Col. Lee Archer had this to say about Noel Parrish: "I flew General Parrish a number of times to meetings in Washington. He would go up there and on the flight back you could almost feel the depression in the man. He had gone up there and gotten beaten up—literally—by the top brass about what was going on at his base. I suppose that it was not such a failure as they had hoped it would be. It is my personal opinion that he ended up with only one star because he defied the system. He was a man of immense courage and honesty. I loved the guy."

In February 1943, Parrish addressed the following words to the 99th: "All of you know that any time you may be sent into an active combat zone. Since none of us can foretell the day of your departure from the station and the command, I shall take this occasion to speak a few words of farewell. You are fighting men now. You have made the team. Your future is now being handed into your own hands. No one knows what you will do with your future as fighting men, you yourselves do not know. Your future, good or bad, will depend largely on how determined you are not to give satisfaction to those who would like to see you fail." Parrish understood!

When the first class moved on to Tuskegee Army Air Field for basic flying school in November 1941, the new field was hardly ready. Because of construction delays and poor weather conditions, only one runway was ready for use at the 1,600-acre facility and the route from the temporary housing for the student

Lee A. "Buddy" Archer

LT. LEE A. ARCHER 302ND. FS., 332ND. FG., AND "INA THE MACON BELLE"

Lee A. "Buddy" Archer, Jr., enlisted in the Army in 1941 and became an aviation cadet at Tuskegee in 1942. In keeping with completing the cadet course with honors, Archer went on to achieve honors in military and business worlds. He flew with the 302nd Fighter Squadron and earned medals for bravery, including the Distinguished Flying Cross, the Air Medal with 12 Oak Leaf Clusters, and the Legion of Merit. Accepting a commission in the Regular Air Force at the end of World War II, he served in responsible positions; the last prior to retiring as a Colonel in 1970 was as Deputy Commander of Bases, USAF Southern Command. In 1970, he took a position with General Foods, Inc., and served as Vice President for Urban Affairs.

No blacks achieved "Ace" status during World War II, although the official publication of the Tuskegee Airmen stated, "…the nation's only black 'ace,' Lee Archer was credited with shooting down more German planes than any other black flyer." Artist Roy La Grone painted five "kills" on his painting of the P-51, Ina the Macon Belle, named in honor of Archer's wife. "I was actually credited with four and a half," said Archer. "Forty years later, I was offered an opportunity to review the records and almost guaranteed that I would get it [Ace status]. I refused that offer. It had been too many years in which they hadn't given it to me and that was that. I have been content."

The mission in which Archer was credited with a half-kill was the first mission in which he downed an enemy. He was flying on Wendell Pruitt's wing. "I saw a Bf-109 and told Pruitt, but he didn't turn so I left him and went to chase the enemy. Freddie Hutchins also had seen it, and he split off, too. We hadn't been overseas too long and we were not really in control. I got onto the tail of the Messerschmitt and Hutchins was almost on my wing, not really in formation. I took a couple of shots. I could see the left wing disintegrate and he started down. The pilot never got out of the airplane. But, as he was going down, Hutchins took a shot at it for the hell of it. The Air Corps—the powers that be—didn't know to whom to credit the kill. Interpretations

pilots to the flight line was never the same. Class 42-C may have been the only flight school class in AAC history to be chauffeured to and from training! Sergeant "Pompy" Hawkins never took the same route twice in the old Ford he drove and the cadets took bets as to which way he would go and how long it would take.

Captain Harold Sawyer, of Columbus, Ohio, a member of Class 43-D, described a typical day at TAAF this way: "We had to be up at 6, breakfast at 7 or 7:30. We only had one dining hall, so it was crowded. You had to get in and get out. Calisthenics after

breakfast, then ground school. In the first part of basic, we flew in the mornings and in the last part we switched and flew in the afternoons. It was the same way in advanced. After we left the flight line, about 3 or 3:30 in afternoon, we had free time between then and dinner, then you hit the books for study. There was lights out at a certain time. Until you got up to advanced, anyone who was ahead of you was constantly hazing you—"Dummy, do this," and, 'Dummy, do that.' You didn't have more than an hour to yourself. It was a full day. You were glad to go to bed at night."

and decisions were made at 15th Air Force. It may be that they were being 'nice' to give us both a half a kill. It was my first, so there was no conscious decision, 'No, he isn't going to be made an ace.' They had no idea there would be any more. Freddie, on a number of occasions, tried to give me that kill."

Archer recalled another exciting day—12 October 1944—that almost ended in disaster. The mission was bomber escort to Blechhammer during which nine enemy fighters were downed in a vicious dogfight. Archer, top scorer with three, was followed by Pruitt with two. (Archer and Pruitt were called, "The Gruesome Twosome!") Archer said, "With five victories, Pruitt and I flew home feeling good! We buzzed the field, then decided to perform a victory roll. Most pilots made a slow roll to the left. Pruitt was left-handed and did his rolls to the right, but I never gave it any thought. I followed Pruitt in, almost on the deck and, as he rolled to the right, I rolled to the left! While I was inverted, my plane slid under Pruitt's! As I fell out of the roll, my plane's wing missed the ground by only a few inches! I still consider myself one of the luckiest pilots in the world."

Archer explained the Tuskegee Experience this way: "We talk about special missions: the invasion of southern France, the mission to Greece when they strung cables across the revetments (those that strafed had to watch that a wing wouldn't come off!), Anzio beach, throwing bombs in the windows of casinos. But, what we talk about most is how we were treated! We conformed to the Army Air Corps and the expectations placed upon us, but we were segregated on Tuskegee Army Air Field! We could get arrested for trying to enter an officer's club! When I'm asked, 'Why did you do it?' I can't fully explain why; but it is my country. My grandfather was born here. Under a system that I consider reprehensible, I flew 159 combat missions and then turned around and went to Korea. It is my country and I will fight for it."

A true hero, Col. Lee Archer has distinguished himself in every way.

Ground school classes were conducted in temporary buildings without partitions and concentration was extremely difficult. Divided into three classes of two students each, the cadets found that they could almost take notes on more than one class at a time. But, despite the difficulties, flying and ground training progressed on schedule and the student pilots became especially adept at making cross wind landings on the single runway. The change from black civilian instructors at Moton Field to white military pilots at TAAF was not quite as dramatic as the cadets had feared. Their new instructors were all volunteers who were as anxious as Colonel Parrish to make sure that the cadets were trained in just the same ways and just as well as their white counterparts. They were extremely demanding and often very vocal.

The pressure was fierce. Harold Sawyer remembered the tension. He said, "They were NOT going to let but so many graduate. Period. That was it. It was frustrating, because some of the guys would go clear up to the night before graduation and then be washed out. You were on a tightrope for eight to nine months. You never *knew*. As you went onto the flight line for your hour, you never knew whether you were going to be coming back the next day or not!"

But, the instructors, in BT-13s for basic and AT-6s in advanced, were competent and conscientious. Three who worked especially hard in the early days at TAAF were Capt. Gabe Hawkins, basic training director; Capt. Robert Long, Director of Advanced Training, who was known affectionately and not so affectionately as "Mother" Long; and, Maj. Donald McPherson, "Black Barney," from his perpetual five o'clock shadow, Director of Fighter Training. None of the three was fully appreciated until actual combat flying proved the wisdom of their instruction and their method of imparting it.

In addition to fair and conscientious training given by their instructors, Tuskegee cadets had good cooperation from the neighboring basic school at Gunter Field and advanced school at Maxwell, both in Montgomery, Alabama, about 40 miles away, and the Air Corps Gunnery School at Eglin Field, in the Florida "panhandle." They had full use of the auxiliary fields, of the night flying facilities and even, when required, their aircraft.

Although there were frequent and minor individual problems with racial overtones, predicted major incidents were prevented by self restraint all around. The closer the men worked together, the greater the understanding and the common desire to get the job done right. Even problems with the surrounding communities proved to be relatively minimal, although, on occasion, an irate farmer would take a shot or two at a low-flying Army airplane. Whether the shots or the "buzzing" came first wasn't always clear. One thing that contributed to the cordiality of the relationship with people of the town of Tuskegee and the surrounding countryside was the fact that cadets and enlisted technicians were kept very close to the base by their schedule demands. Their presence in the local communities was restricted and their social life was focused at the base and at Tuskegee Institute.

Party time at Tuskegee Army Air Field brought Tuskegee Institute coeds and the airmen together—but not too close together. The "six-inch" rule prevailed. *Office of Air Force History, Maxwell AFB, AL*

Field, named for Booker T. Washington's successor as President of Tuskegee Institute, and Tuskegee Army Air Field, but both were subject to delays and conditions were somewhat chaotic.

It was into this environment that Colonel Kimble came as the new Commander of TAAF—a very different type officer than the straightforward Major Ellison. Kimble went to great pains to practice what he believed to be "the Southern way of life." He enforced strict segregation at TAAF and catered to the white minority in the surrounding countryside. His leadership impact on trainee morale seemed to have been of little consequence to him; he believed that the authorities in Washington would have been happiest to see the experiment fail completely. Only his sense of duty as a soldier and the guidance of General Weaver at SEAACTC prevented him from killing the program himself. To his credit, he was an excellent organizer and, in his year as Commander, tremendous progress was made in bringing a high degree of order to the new base and its organizations.

As the first class came to graduation on 7 March 1942 and as subsequent classes continued their training, Tuskegee Army Air Field began to show the reality of what had been a most ambitious plan. Hangars, repair shops, classrooms, barracks, administrative buildings, dining and recreation halls, and health care facilities were nearing completion and all landings didn't have to be made on the same runway.

Graduating cadets found themselves under escalating pressure. In addition to that brought to bear by an Army that still wasn't convinced that black pilots could fly effectively in combat, there were demands to succeed coming from black leaders, from the black press, and from the black community at large. By far, the greatest demand came from themselves.

As Dr. Rose expressed it in *Lonely Eagles*: "It was probably difficult enough to fly, without feeling that 13,000,000 black Americans were depending upon them as a source of inspiration."

Lieutenant Mac Ross brought great pressure upon himself soon after graduation and during his initial days of transition training in the P-40F. While flying in formation with his wingmen Custis, DeBow, and Roberts, smoke began pouring from his engine cowling. The formation loosened somewhat, began a descent from 6,000 feet to 3,000, and headed back to the base. By the time they reached 3,000 feet, the smoke was worse and Ross' companions persuaded him to bail out. Landing safely in a cotton field, he was relieved to be alive, but worried that he had ruined the program. Although he was immedi-

Entertainment at Tuskegee was outstanding and it helped offset the grim segregated environment. Among the stars who visited TAAF were Louis Armstrong, Joe Louis and Lena Horne. The base boasted excellent sports programs, informal training courses and musical groups like the very talented Imperial Kings of Rhythm. The aviators' success in dating Tuskegee coeds was an irritant to the men of Tuskegee and the Institute's "six inch rule" that kept couples that far apart was frustrating to the fliers. But it all seemed to work out. Noel Parrish was quoted, after the war, as saying: "Tuskegee Air Field operated for five years with many hundreds of Negro women and several white women. During this time almost every conceivable interracial complication arose except one—sex."

1942

As America struggled to recover from the effects of the devastating attack on Pearl Harbor and the Declaration of War, the environment at Tuskegee became even more stressful. Civilian Pilot Training Program courses became war-related within a week of 7 December.

Those washed out of military pilot training were retained at TAAF as Privates, although they had little or nothing to do and were a clear detriment to morale. Construction progressed at Moton

ately and completely cleared of any pilot error, all he could think was: "I've wrecked a ship worth thousands of dollars. Maybe they'll start saying that Negroes can't fly after all!"

His fears weren't realized and the first class continued its transition to combat flying in the P-40. The second and third classes, 42-D and 42-E, completed advanced training and were awarded their pilot wings at 4 1/2 week intervals. In the second class, there were just three graduates—Lieutenants Sydney Brooks, Charles Dryden, and Clarence Jamison—and four in Class 42-E: Lieutenants James Knighten, George Knox, Lee Rayford, and Sherman White. With the need for 33 pilots to fully staff the 99th and only twelve trained, it would be a long time before the Squadron was ready for deployment.

Yet the War Department and the Air Corps continued to resist urging from Judge Hastie, other black leaders, and the black press to abandon quotas and to train more of the qualified blacks who had been accepted. The problem was obvious in the numbers of letters received by the NAACP, by sympathetic Congressmen, and by the President himself from frustrated pilot candidates whose applications had been accepted months and years earlier and who still hadn't been called. It had been made even more obvious by activation of the 100th Pursuit Squadron at Tuskegee on 19 February 1942. The burning questions were: Where are the people to staff these units coming from and when?

Captain Parrish attributed the relatively high "washout" rate among the first classes to a lack of familiarity with aviation among the cadets. He proposed that future cadets come exclusively from graduates of the secondary (advanced) CPTP course. His recommendation was accepted at SEAACTC but rejected at AAC Headquarters because this would result in selection criteria for blacks being different from those for whites. By mid-1942 the proposal became moot as CPTP was phased out in favor of the CAA's War Training Service (WTS) for which civilians were ineligible.

A look back at these two highly effective programs shows not only tremendous numbers of trained pilots but real steps forward for blacks and women in aviation. Those steps, however, were severely limited in impact by continuing quotas on black military pilot training spaces and by the exclusion of women from military flying. Mr. Washington, Tuskegee's director of the CPT program, had, from the beginning, regarded the program as a "demonstration" rather than as an "experiment." He believed that, given the opportunity, blacks could learn to fly in the same proportion as whites. The CPT program proved him correct!

An experience of Herman "Ace" Lawson is just one example of the impact of Air Corps attitudes toward the training of blacks as pilots. Having earned his Private Pilot's license in CPTP at Fresno State College in California, Lawson went with his white classmates to inquire about making application for the cadet program. The white Major with whom he spoke told him: "Get the hell out of here, boy. The Army isn't training night fighters." When he learned that blacks were being accepted for training at Tuskegee, Ace immediately applied. He waited for five months, then wrote to his Congressman, his Senator, and, finally, the President. After another two months had elapsed, he was notified that he had been accepted.

And so the problems continued: high washout rates, with nowhere to go, brought about mainly by lack of aviation familiarity, and low quotas for black trainees brought about by the Army Air Corps' indifference, if not hostility. Each step the AAC took resulted from continuing pressures in the political arena and in the press, not from any perception of a need for manpower to win the war.

Lieutenant General Benjamin O. Davis, Jr., assumed command of the 99th Fighter Squadron from Capt. Harold Maddux on 24 August 1942. In addition to a problem of personnel shortages, he faced another that was equally severe. As time went on and as additional classes of pilots graduated, transition training continued for those in earlier graduating classes. Maintaining discipline and morale became increasingly difficult as months dragged on, as training exercises were repeated over and over again, and as other units went off to fight for their country.

In December, he spoke to the Squadron at a Field Day program. He said, "The success of the combat unit will prove to be the opening wedge for the air minded youths who aspire to the field of Aviation. The records, so far, by the cadets at Tuskegee Army Flying School do not compare unfavorably with those records made by cadets at other fields. My greatest desire is to lead this squadron to victory against the enemy."

1943

Nine months elapsed after the graduation of Class 42-C at Tuskegee. The 100th Fighter Squadron had been activated in February of the previous year with Mac Ross as Squadron Commander; the 332nd Fighter Group was activated in October under the command of Lt. Col. Samuel Westbrook; and Classes 42-D through 42-K had earned their wings. Enlisted technical and administrative specialists completed their elongated training elsewhere and came to

Roscoe C. Brown Jr., Ph. D.

LT. ROSCOE BROWNE 100TH. FS., 332ND. FG., ONE OF THE FIRST ARMY AIR CORP PILOTS TO SHOOT DOWN A GERMAN JET ME262 ✠ MARCH 24,1945

Roscoe Brown, Ph.D., grew up in Washington, DC, the son of the originator of the Negro health movement and in charge of public health for the nation's blacks. A graduate of Dunbar High School, Brown recalled being raised in a climate that stressed the importance of education, achievement, and a commitment to addressing social problems. Also a Captain, Roscoe Brown commanded the 100th Fighter Squadron, the first of the original 332nd Fighter Group to become active. Brown commanded the squadron until it returned to the States in October 1945.

In February 1945, the Group destroyed five enemy aircraft on the ground, seven locomotives, five box cars, and one flat car. Brown was credited with one locomotive. He admitted that his most frightening combat experience was during the strafing attack of a train. He said, "I was leading the attack and decided to get low so that we wouldn't be easy targets. The train had a flatcar that dropped its sides to reveal guns which started firing. Unfortunately, I got so low that I hit the train with one of my wings. I grabbed the control stick with both hands and flew back with half a wing angled up in the air."

On 24 March 1945, B. O. Davis, Jr. led the 332nd on one of the longest missions ever attempted by the 15th Air Force (1,600 miles). The pilots were briefed to bomb the Daimler Benz Tank Works in Berlin, but the mission was actually designed as a diversionary effort to draw off German fighters based in central Germany which might otherwise have been committed against the airborne landings north of the Ruhr. It was planned that the 332nd would relieve a P-38 Fighter Group at 11 o'clock over Brux and carry the bombers to the outskirts of Berlin. At this time, the 332nd was to be relieved by another P-51 group that would escort the bombers over the target.

Upon arrival at the relief point, the 332nd was instructed to continue over the target because the relief group was late. Over the target, the 332nd encountered the new German jet planes for the first time. In the ensuing battle, pilots of the Group, flying piston-powered aircraft, destroyed three of the jets, that had to slow in order to attack the U.S. bombers. Captain Roscoe Brown, Lieutenant Earle Lane, and Flight Officer Charles V. Brantley were credited with the victories. Brantley wrote later, "I flew 42 combat missions as a fighter pilot…and was credited with downing one of the first jet type aircraft in aerial combat. Flying I liked and I considered myself a real tiger."

Brown, a flight leader in the 100th at the time, said, "I sighted a formation of four Me-262s under the bombers at about 24,000 feet. They were below me going north. I peeled down on them toward their rear, but almost immediately I saw a lone Me-262 at 24,000 feet, climbing at 90 degrees to me and 2,500 feet from me. I pulled up at him in a 15-degree climb and fired three long bursts…. Almost immediately, the pilot bailed out. I saw flames burst from the jet orifices of the enemy aircraft."

For successfully escorting the bombers, the Group was awarded the Distinguished Unit Citation, then went on to further outstanding victories. On a strafing mission, 31 March 1945, a flight of enemy aircraft mixed in a furious dogfight with members of the Group, who shot down an astounding thirteen of the enemy planes without losing a plane. In that aerial skirmish, Lt. Roscoe Brown was credited with downing one FW-190.

Dr. Brown, past President of Bronx Community College of City University of New York, a position that he held for 16 years, received his Bachelor's degree from Springfield College, MA. He holds a doctorate from New York University and has served as a faculty member at West Virginia State College and as a full professor at NYU's School of Education. Listed in Who's Who in America, Dr. Brown started NYU's Institute for Afro-American Studies which he directed for 7 years. An acclaimed civic leader, Dr. Brown was president of the prestigious "100 Black Men" of NYC in 1986. For his wartime exploits, he was awarded the Distinguished Flying Cross and Air Medal with eight Oak Leaf Clusters.

Tent City, Tuskegee Army Air Field, was often ankle deep in mud; good practice for the Italian winters that followed. *Office of Air Force History, Maxwell AFB, AL*

Tuskegee to form the support elements of the 99th. The 83rd Fighter Control Squadron and the 689th Signal Air Warning Company completed their training and, like the 99th, were engaged in a seemingly endless round of combat preparation exercises.

Gene Carter described the situation: "We were combat operational in October of '42 and we didn't get to combat until April of '43! All that while we were at Tuskegee accumulating flying time. Every other month you were down at Eglin Air Force Base going through Tactical training—gunnery, dive bombing, strafing."

And there was still no definitive word on overseas deployment. Rumors were rampant and false alarms kept the men in a state of virtual confinement. They even flew on Christmas and New Year's Day. Mildred Carter described the frustrations of indecision and expressed the sentiments of all when she said, "We just kept saying 'Good-bye,' over and over and over."

Lieutenant Colonel Parrish succeeded Colonel Kimble as TAAF Commander. He and Colonel Davis cooperated to do all that they could to maintain a fighting edge and a reasonable level of morale. Feedback from the once highly reluctant War Department was positive and an announcement was made to the effect that it expected the deployment of the first Negro Squadron overseas shortly.

The Adjutant General of the Army said: "From results so far obtained, it is believed that the Squadron will give an excellent account of itself in combat and that it will be a credit to its race and to Americans everywhere."

Even the training quotas had been quietly expanded. Classes beginning with 43-C were substantially larger. Combat training for pilots and ground crews intensified. Secretary of War Stimson paid a visit to Tuskegee in February. After being shown what the Squadron was doing, he commented that the outfit looked as good as any he had seen. The possibility of overseas movement became a probability, and tension, already high, mounted.

On April 1st, the 99th Fighter Squadron finally prepared for departure. Although their move was supposed to be confidential, news of impending transfer to combat leaked out. A crowd of well-wishers jammed Chehaw station when the train pulled out for Camp Shanks, New York.

Overseas processing took ten days, during which the men were able to pay brief visits to the City on the Hudson. Some saw family and friends. Colonel Davis, who would be the senior officer on board the troop ship which contained white troops as well—another first for a black officer—was designated Executive Officer for the long voyage ahead. He chose his staff, many from among the officers of the 99th, and oversaw final preparations.

USS Mariposa sailed from New York harbor before the sun was up on April 15th. Four thousand officers and men aboard this former luxury liner strained the ship's capacity, especially when sea-sickness struck so many the first day out. As they got their sea legs, the stomachs quieted down and the troops settled in for what proved to be a rather uneventful crossing.

Africa was sighted on the morning of the 24th and, by late afternoon, the ship entered the harbor at Casablanca, French Morocco. Disembarkation was a long and frustrating process, as one unit at a time went down the gangplank.

Once off the ship, the 400 officers and men of the 99th were driven through town en route to their temporary bivouac area, where they set up camp, rested for the night, and remained until they were satisfactorily equipped for combat. When Colonel Davis was convinced that they were prepared for the weeks ahead, the 99th set off for its first relatively "permanent" home overseas—OuedN'ja, near the city of Fez and between Casablanca and Tangier.

It took seventeen hours to go the 150-mile distance. Once there, they began a period Colonel Davis described as "...probably the most pleasant we were to enjoy during our overseas tour." In a training status under the North Africa Training Command, the 99th was stationed with the 27th Fighter Bomber Group commanded by Lt. Col. John Stevenson, whom Colonel Davis had known at West Point. Informal "dog fights" between the 99th's brand new P-40Ls and the

27th's A-36s helped the pilots of the 99th learn some important aerial tricks and experienced warriors like Lt. Col. Phil Cochran (the pilot that inspired "Flip Corkin" of comic strip fame) and Majors Fachler and Keyes provided knowledgeable battle tactics and insight. Squadron members had considerable time in the aircraft; what they lacked was experience in combat.

Although combat was their focus, diversion from the heavy training schedule brought together members of the two organizations—one black and the other white—in sports contests and men from the two units visited Fez regularly and without incident. Josephine Baker, a popular American black expatriate singer of the 1920s, entertained the troops and introduced their officers to prominent French and Arab families. In addition to the pleasant conditions and learning opportunities, the 99th ferried in its own new airplanes from Casablanca—27 P-40Ls equipped with Packard-built Merlin V-1650 engines. Accustomed to cast-offs, the pilots and mechanics were delighted to fly and to maintain their first new aircraft.

After a month of relatively intensive combat training, the Squadron moved, by air and by train, to Fordjouna, on the Cape Bon peninsula in Tunisia. The 99th was attached to, but not a real part of, the 33rd Fighter Group, commanded by Col. William "Spike" Momyer, part of General Cannon's Northwest African Tactical Air Force. Lieutenant Colonel Cochran, known as "Mr. P-40" because of his daring exploits in the North African Campaign, continued to provide pilots of the 99th with practical combat training and declared, "These guys are a collection of natural-born dive bombers." Their training was essentially over; it was time to go to war!

Pantelleria

On 2 June 1943, the men of the 99th flew into combat for their history-making debut. Lieutenants William Campbell and Charles Hall, flying as wingmen with the 33rd Fighter Group's experienced pilots, flew on that first mission, a strafing and dive bombing mission to the fortified island of Pantelleria, about halfway between the Cape Bon Peninsula and Sicily.

Bill Campbell, widely credited with being the first Tuskegee Airman to fly in combat, modifies the story with characteristic modesty, saying that, as they released their bombs, Charlie Hall was only seconds behind! Lieutenants Clarence Jamison and "Little Flower" Wiley flew the next Pantelleria mission. All returned safely to Fordjouna after essentially routine missions with no enemy aircraft sighted. Of the unit's readiness for combat, Colonel Davis said, "I personal-

Erwin B. Lawrence

Captain Erwin B. Lawrence, Jr., Operations Officer, was reported missing from a strafing mission to the Athens Tatoi Airdrome, Greece, on 4 October 1944. His loss was sorely felt throughout the Group.

Lawrence, a pilot from the original 99th Fighter Squadron, was born in Cleveland, Ohio, on 31 May 1919. Commissioned a 2nd Lieutenant, Lawrence entered the 99th Fighter Squadron on 3 July 1942.

Tasting combat with the squadron in June of 1943, Lawrence embarked on a remarkable—though regretfully short-lived—career. He participated in close to one hundred dive bombing, strafing, patrol, and escort missions throughout the Pantellerian, Sicilian, and Italian Campaigns. He was credited with a probable kill over Anzio, Italy, on 27 January 1944—a Folke Wulf-190.

Captain Lawrence assumed command of the 99th Fighter Squadron as successor to George "Spanky" Roberts in April of 1944. He was held in high respect and relieved of his command only when Major Roberts returned from the United States the following September. At that time, Captain Lawrence became the Operations Officer for the squadron.

Lawrence led pilots of the 332nd Fighter Group on the fateful mission to Greece—the targets: Eleusis, Kolamoi, Megara, and Athens Tatoi Airdromes. Lawrence dove toward Tatoi with other members of his element following in string formation. The Germans opened fire and, as Captain Lawrence prepared to pull out, a barrage of machine gun fire raked his aircraft. Captain Lawrence spun into the ground, a casualty whose loss was keenly felt.

Captain Lawrence was awarded the Air Medal with three Bronze Oak Leaf Clusters. He is fondly remembered by all who knew him.

John William Mosley

3ND. LT. "BIG" JOHN MOSLEY #77 BOMB GROUP

Born in Denver, John W. Mosley was educated in the public schools of Colorado, earning a place in the National Honor Society. He was awarded a Merit Scholarship to attend Colorado State University, where, as a scholar/athlete, Mosley excelled in football (All-Conference All American, Most Valuable Player) and wrestling (All Conference—four years). He was elected vice president of his college senior class.

A welcomed volunteer for the U.S. Army Flying Training School, Tuskegee, Mosley carried his outstanding qualities into flight training. He remained at Tuskegee from 1943 through 1945.

Leaving the military after World War II, Mosley obtained his Master's degree from Denver University. With a major in social work, he was employed with the YMCA programs until 1952, at which time he reentered the U.S. Air Force. He served in Wiesbaden, Germany; Greensboro, North Carolina; Pasadena, California; Lackland Air Force Base, Texas; and Lowry Air Force Base, Colorado, prior to being stationed with the 388th Tactical Fighter/Bomber Wing in Korat, Thailand. In the latter assignment, Mosley served as Chief of Combat Operations, Plans and Analysis. Before he retired, Lieutenant Colonel Mosley accrued over 9,000 flying hours in single- and multi-engine aircraft. A Command Pilot,

his decorations include the Bronze Star, the Air Force Commendation Medal, and an Outstanding Service Award.

Mosley and his wife, the former Edna Wilson to whom he was married in 1945, have a daughter and three sons. His innumerable awards include being named as Citizen of the Year in Denver, in Who's Who in Black America, and Outstanding Black in Colorado History. He wrote, "I assisted my wife, Edna, who chaired the entire project, in developing and implementing the Royal Pacific Cultural Exchange with the Republic of Singapore, a joint venture between Sister Cities International and United Airlines. As a member of the Board of Directors for Goodwill Industries, Inc., of Denver, I have been responsible for increased emphasis on rehabilitation programs for severely handicapped minorities, among other important activities. I am simply reminded daily 'that it is not what one has done in the past, but what one is doing now' that is important. My wife and I work together on economic development, family resource centers, health care, and a number of other worthwhile and important ventures."

"Big John" Mosley—another outstanding Tuskegee Airman—has carried the excellence with which he was inspired throughout his life.

ly believe that no unit in this war has gone into combat better trained or better equipped than the 99th." What was missing was combat experience; that took many missions to acquire.

For a week, the men of the 99th participated in attacks on the island—never seeing enemy aircraft. They were now flying as an independent unit with no 33rd Fighter Group pilots to provide direction and example.

Finally, on 9 June, German aircraft attacked a 99th flight of six P-40Ls, led by Lt. Charles Dryden. The flight of 12 Focke-Wulf-190s and Messerschmitt Bf-109s were escorting 18 bombers in a raid on the Allied force that was preparing to invade Pantelleria. In the ensuing dogfight, Lt. Rayford's aircraft was damaged by two German fighters; but they were driven off by Lieutenant Spann Watson's machine guns. Lieutenant Willie Ashley very nearly logged a "kill" to become the first black pilot to score an aerial victory in World War II, but that had to wait until another day, another fight. The Germans turned for Sicily and the Tuskegee Airmen returned to base.

Northwest African Air Force air attacks and shelling by British naval forces continued relentlessly for thirteen days, bringing about a surrender of the Italian garrison on Pantelleria on 11 June 1943. The 99th flew 16 sorties a day in this Air Combat Campaign. They participated in destroying the enemy's will to resist before an assault force had even gone ashore—an unprecedented coup. Other Italian islands between Tunisia and Sicily, Lampedusa and Limosa, succumbed to Allied aerial attacks and naval gunfire on 12 and 13 June. Sea lanes between North Africa and Sicily now belonged to the Allies and the way was paved, first for the invasion of Sicily and then for the assault on the European mainland.

Sicily

Even before the surrender of Italian island garrisons, aircraft from Northwest African Air Forces, based in Algeria, and the Ninth Air Force, flying from Libya, were regularly attacking German and Italian forces at Sicilian airfields and in the harbor at Messina on the northeastern coast. They attacked troop concentrations, roads, and defensive positions and flew similar attack missions against the island of Sardinia and against the Italian mainland. By early July, these attacks, coupled with very aggressive air-to-air combat, resulted in Allied air superiority in the skies over Sicily. Unchallenged reconnaissance flights provided invasion planners with invaluable data and photographs of Axis positions and strengths.

The 99th, re-assigned to the 324th Fighter Group for logistics and administrative support, moved on 29 June from Fordjouna to the tip of the Cape Bon Peninsula and flew its first mission to Sicily on 1 July—a medium bomber escort mission that proved successful but lacked challenge.

An escort mission the next day more than made up for that lack! Six P-40s of the 99th were charged with providing escort for 16 B-25s attacking Castelvetrano. After the bombers released their ammunition, they flushed German FW-190s that streaked like angry bees to avenge the onslaught. As two of the enemy pulled into firing position behind the B-25s, Lt. Charles Hall wheeled between the bombers and their attackers. He fired a long burst that caught the second Focke-Wulf and sent him down. A Tuskegee Airman had scored his first aerial victory! The entire squadron rejoiced!

Lieutenant Hall wasn't alone in the sky. Lieutenants Dryden and Knighten were also engaged, first by two Bf-109s and then by two FW-190s. As the hunter becomes the hunted, Dryden and his wingman chased the Messerschmitts and, in turn, were attacked by the Focke-Wulfs. Breaking off their attack, the P-40 pilots desperately maneuvered to distance themselves from the enemy's guns. After an extended chase, the Germans turned back. Dryden and Knighten made it home, only to discover the severe damage to Dryden's aircraft!

Hall's victory was greeted with great jubilation. He relished the cold Coca Cola that had been on ice awaiting the Squadron's first kill and he happily received the personal congratulations of Generals Eisenhower, Spaatz, Doolittle, and Air Marshall Cunningham of the RAF. The Squadron's elation was dimmed, however, by the deaths of Lieutenants Sherman White and James McCullin. On takeoff earlier that morning, their aircraft fatally collided!

Harold Sawyer admitted, "This is a hard thing to say, but you couldn't afford to dwell on any fatal accident. You had to take it in stride. When you see a friend go down and the next day you go back up, then you *have* to think about something else. If you dwelled on it, you would be the next one in trouble."

The 99th continued to play an active role in the lead-in to the coming Sicilian invasion. They escorted medium bombers and strafed and bombed enemy targets. Just prior to 10 July invasions by General Patton's U.S. Seventh Army at Gela and Licata on the southwest coast and Gen. Bernard Montgomery's British Eighth Army on the southeastern coast below Syracuse, they provided aerial cover for assault forces in their low-level attack missions against Axis positions. Seaborne assault forces were augmented by aerial assaults by paratroopers and gliders. These assaults, although decimated by friendly naval antiaircraft guns and hampered by high winds, did help to secure the

flanks of the invasion forces. The Allies moved north toward Palermo and Messina.

Patton's Army met with relatively less resistance and, by 23 July, captured Palermo. The British Eighth encountered heavy German resistance around Syracuse and made slower headway in the push toward Messina. Throughout the campaign, the Ninth and Twelfth Air Forces continued to attack enemy targets in Sicily, in Italy, and in the narrow sea lane between island and mainland. In addition to reducing enemy resistance, they sought to limit reinforcement/resupply and escape.

The 99th completed its attachment to the 324th Fighter Group on the day of the invasion. In eleven days, the Squadron flew 175 sorties—principally ground attack missions—boasted one confirmed kill and two probables. For a little more than a week, the 99th was attached to and provided air support for General Montgomery's Army. On 19 July, the Squadron was again satellited on Colonel Momyer's 33rd Fighter Group and moved to an airfield at Licata, demonstrating the mobility of a single squadron—and its proximity to the front.

Of relocation, Colonel Carter said, "With a single squadron, it was easier to come in with a bulldozer and knock out a couple thousand feet and put down some PSP, metal runway and tent flooring. Each man had his one-man pup tent. You'd go into the olive grove, dig your trench, put your tent up and your bedroll down and that was your house. In Sicily—Licata—there were times that we would have to go out over the Mediterranean to climb to altitude to avoid antiaircraft fire. That's how close we operated to the bomb line."

Welcome relief in the form of replacement pilots joined the 99th on 23 July—Lieutenants Howard Baugh, John Gibson, John Morgan, and Edward Toppins. In contrast, Lieutenants Ace Lawson and Clinton "Beau" Mills, who had left Oscoda, Michigan, where they had been training with the newly constituted 332nd Fighter Group, were stranded in Brazil for three weeks. Even when they finally showed up at the end of July, Colonel Davis still hadn't seen their transfer orders. Nevertheless, they, too, were welcomed with open arms and put swiftly to work.

The Sicilian Campaign continued. Patton's troops pushed toward Palermo. Montgomery continued a slow advance toward Messina and Axis forces were relentlessly forced toward that critical port city from which they hoped to make their escape. The 99th continued its dive-bombing and strafing of retreating forces; it patrolled the skies above the battle lines and performed armed reconnaissance missions on a regular basis. It did whatever was asked of it, but there were no more aerial encounters with the enemy's air forces.

Another tragic accident occurred on 11 August. In a mid-air collision, Lt. Graham Mitchell's P-40 struck a plane flown by Lt. Samuel Bruce. Mitchell was killed when his plane crashed into the ground. Although he was only at 1500 feet, Lt. Bruce parachuted to safety.

The Axis forces essentially gave up on Sicily during the first part of August and, although they were under continuous attack from the air, managed to get a substantial number of troops and considerable materiel across the narrow Straits of Messina to the Italian mainland. On 17 August, the U.S. Third Division arrived in Messina; with elements of the British Eighth right behind. Together they "mopped up" the remaining enemy forces and the island was secured. Thirty-eight days were required to complete the campaign, which resulted in 100,000 prisoners.

A More Dangerous Enemy

In late August, Colonel Davis was relieved of command of the 99th and transferred back to the U. S. to command the 332nd Fighter Group, then in training at Selfridge Field and Oscoda AAF in Michigan. Captain George "Spanky" Roberts, Operations Officer of the 99th, took over the Squadron. Both were unaware that, while the 99th was engaged with the Pantelleria and Sicily Campaigns, Army Air Corps brass was becoming the real, and potentially, more dangerous enemy!

War Correspondent Ernie Pyle later wrote: "Their job was to dive bomb, and not get caught in a fight. The 99th was very successful at this, and that's the way it should be," but pilots of the Squadron were criticized for their failure to down more enemy aircraft. Roberts was visited by Chief of Staff Gen. Henry "Hap" Arnold shortly after he assumed command. Spanky said of the visit: "I never felt so bad in all my life. It was extremely difficult for me as he stood there discrediting and maligning the men who were doing their best under the circumstances. Whether right or wrong in his assessment of the unit, his manner was not what should be expected from the Chief of the Air Corps."

The press at home, supported by partial reports and half-truths from the field, began reporting a negative view of the 99th. *Time* magazine's September 20, 1943, edition included an article headed "Experiment Proved?" which was based partially on a brief interview with the returning Colonel Davis and partially on rumors and tidbits of information leaking back from the front.

Class 43-A, the first in the new year at Tuskegee Army Air Field, went into combat readiness training with the 99th and newly formed 100th Fighter Squadrons. *U.S. Air Force Museum*

An advanced training instructor points out significant check points prior to cross country flight training. *Official USAF Photo, Roy E. La Grone*

Among other things, the article said: "It [the 99th] has apparently seen little action compared to many other units" and "unofficial reports from the Mediterranean Theater have suggested that the top air command was not altogether satisfied with the squadron's performance. There is said to be a plan to attach it to the Coastal Air Command, in which it would be assigned to routine convoy cover."

An article in the September 30th edition of the *New York Daily News* added fuel to the fire when it stated, "Although thousands of Negro soldiers have been drafted, only one outfit, the 99th Air Squadron, has been engaged in actual combat. This had brief action in Africa but has since broken up and its members returned to the United States for training purposes."

As the reaction to these articles developed, the official reports from the 99th's chain of command arrived in Washington. Colonel Momyer, commander of the 33rd Fighter Group, to which the 99th had twice been attached, wrote the following to his commander: "Based on the performance of the 99th Fighter Squadron to date, it is my opinion that they are not of the fighting caliber of any squadron in this group. They have failed to display the aggressiveness and daring for combat that are necessary to a first class fighting organization. It may be expected that we will get less work and less operational time out of the 99th Fighter Squadron than any squadron in this group."

Major General Edwin House, commander of the XII Air Support Command and Colonel Momyer's boss, explained his understanding of the situation this way, "the Negro type has not the proper reflexes to make a first class fighter pilot." House recommended sending the 99th back to Africa on coastal patrol and convoy protection duties and re-equipping it with P-39s in order to release their P-40s to a more "aggressive" squadron.

Major General John Cannon, House's commander as C.O. of the Northwest African Tactical Air Force, endorsed the Momyer and House reports to Lt. Gen. Carl Spaatz, commander of the North African Air Force, in a way that fully supported them. Spaatz' endorsement to General Arnold was somewhat more cautious, noting that the 99th had shown excellent ground discipline and ability to execute orders and that no unit had a better training background.

These negative reports fit well with Arnold's personal views of the black airmen. However, he did order that studies be made of questions regarding the quality of the black pilots. The Air Staff reported that the 99th had required eight months of combat training, compared to three months for comparable white squadrons, without further explanation of the situation. The Operations Division of the War Department cited a report from General Eisenhower that, through early August 1943, the 99th had creditably carried out difficult strafing missions in Sicily.

The entire package on the 99th Fighter Squadron was put before the Advisory Committee on Negro Troop Policies. Colonel Parrish and Colonel Davis testified before the Committee. Both reported that while mistakes had been made in the early combat experience of the Squadron, the experience was growing and performance improving. Parrish especially noted the negative impact of seg-

regation on the 99th's ability to benefit from the combat experience of white units with which it was associated. In characteristic fashion, Davis directly addressed the points raised in criticism of the Squadron and stated that; "I carried out my mission. If given a mission to bomb a target, I went ahead and bombed it."

Despite a recommendation from the War Department that the 332nd Fighter Group, to include the 99th, be sent to the Mediterranean area to provide a "just and reasonable test" of a large Negro air unit in combat and that plans be made to expedite the formation of a Negro medium bombardment unit, General Arnold was determined to end the combat career of the 99th. He ordered a draft memorandum to President Roosevelt stating that: "It is my considered opinion that our experience with the unit can only lead to the conclusion that the Negro is incapable of profitable employment as a fighter pilot in a forward combat zone." He further proposed the reassignment of the Squadron to a rear defense area.

Fortunately, the memo never got to the President! Colonel Emmett O'Donnell of the Air Staff recommended that the matter be reconsidered. He pointed out, "Every country in this war has had serious trouble in handling disaffected minorities. To recommend at this time any action which would indicate the relative inferiority of the colored race would really be 'asking for it.' …Further, I feel that such a proposal to the President at this time would definitely not be appreciated by him. He would probably interpret it as indicating a serious lack of understanding of the broad problems facing the country," and concluded, "…it might be far better to let the entire matter drop, without any letter to the President." The matter was dropped.

Thus, the Tuskegee Experiment ended where it began—in Washington, DC. The senior leadership of the Army Air Forces had not changed its opinions at all. Political realities prevailed. Three thousand miles away, the 99th Fighter Squadron fought on against the enemies of the United States.

CHAPTER THREE

The Tuskegee Experience

As negative letters about the 99th Fighter Squadron were being written and endorsed, as "studies" were being ordered, and as testimony was taken in Washington, the Tuskegee Airmen pressed on. Largely unaware of the brickbats being thrown at them by their senior leaders, Maj. Spanky Roberts and the men of the 99th kept doing whatever was asked of them; determined, as always, to prove that black men could fight for their homeland as well as anyone. Capability, courage, and a marvelous sense of humor were evident as the war moved to the European mainland.

**99th Fighter Squadron,
September 1943–June 1944**

Two significant events marked the beginning of the 99th's participation in the Italian campaign—finding a physical home and finding an organizational home. On 2 September 1943, the same day that Major Roberts assumed command, an advance party left the airdrome at Licata, Sicily, to set up a base on the Italian mainland. Expected to be a routine movement immediately following the successful Allied landings at Salerno, the advance group came under enemy fire almost as soon as it went ashore near Battapaglia. The 99th's airmen, suddenly infantry soldiers, retreated swiftly and joined the British Tenth Corps where they were hastily equipped with small arms weapons. They were ordered to fire only at enemy foot soldiers and to let German tanks pass by. Luckily, the German counterattack was halted short of the Tenth's lines and they didn't have to comply.

The advance party continued south, moving between German and Allied lines only to discover that the airfield they were to prepare for the

Squadron was still in German hands. Assigned to another field, they moved to it in a fog of exhaustion. There they came under German bombing and strafing attacks for five consecutive days and nights. It wasn't until 17 October 1943, when they were re-united with their squadron mates at Foggia on the Italian east coast, that they felt secure again.

While the advance party was searching for a new home for the P-40s of the 99th, the planes were being flown from an airfield near Barcelona, Sicily, on a variety of routine patrol missions that gave them the distinct feeling of having been left out of the war. Those negative feelings came to a happy end when the Squadron was re-assigned to the 79th Fighter Group, commanded by Col. Earl E. Bates.

Like Colonels Parrish and Davis, Colonel Bates was primarily concerned with two things: the combat proficiency of his Group and the well-being of his people. In every respect, he earned the admiration and affection of the Tuskegee Airmen. Their six-month association with the 79th Fighter Group was truly beneficial; it was, in the words of Col. Gene Carter, "...the first integration in the U.S. Air Forces."

The pilots of the 99th flew as wingmen with the more experienced pilots of the 79th on training and combat missions. They learned more effective takeoff and flight tactics and they truly felt part of the organization because they *were* part of the organization. Missions were laid on without regard to the color of the pilots and the four squadrons flew combat strafing missions together, trained together, and, as limited opportunity was presented, played together.

As the Italian campaign ground on, the weather became a major operational problem. Rain, cold temperatures, and high winds might have been tolerated if there hadn't been the mud!

---

top
Armorers, key members of the team, including Leon Strong, Allen Norton, and John T. Fields, carry belts of .50 calibre machine gun ammunition for awaiting P-40s. *USAF Photo, Roy E. La Grone*

bottom left
Lieutenant John L. Hamilton, first in the 99th Fighter Squadron to be awarded a Purple Heart as the result of wounds sustained in combat. *U.S. Air Force Museum*

bottom right
Lieutenant Roscoe Brown poses by his P-51 Mustang. Lieutenant Brown was one of the very few American pilots to shoot down a German Me-262 jet fighter. *Roy E. La Grone*

A P-40 of the 99th Fighter Squadron in North Africa, first flown by them in combat in the attacks on the Italian island of Pantelleria. *Lt. Col. Gene Carter, USAF Retired*

"Good old Italian mud!" recalled Gene Carter. "If you got off of that PSP steel planking used for runway and taxi revetments, you were in the mud. I didn't worry about getting it off. I wore overshoes. If alerted, I'd just kick them off and jump into my flight suit and go. You couldn't walk to the debriefing tent without getting muddy, so your crew chief would take your boots to and from the aircraft for you."

The 99th Historian stated: "We are up to our necks in mud. Here and there are fallen tents, blown down the night before by the wind. No one soldier has a complete uniform. Every man has a different idea regarding which of his clothing to wear in order to keep warm."

When the weather cleared, combat missions resumed as if they had never stopped. In one stretch, the 99th flew 48 sorties per day in support of the British Seventh Army as it fought to cross the river Sangro. Always nearing the forward edge of the battle area, the 79th Fighter Group relocated to Madna, near Termoli on the east coast, in mid-November. On 22 November, Lieutenant James Wiley became the first 99th pilot to pound a "fifty-mission crush" into his officer's cap. Thanksgiving was celebrated with a traditional feast prepared in the 79th's mess tents.

December's weather was little improved. Bombing and strafing missions were flown as conditions permitted and morale seemed to slip ever lower. Charlie Hall's June 1943 aerial victory remained the only one and ground crewmen erroneously began to doubt the courage and capability

of their pilots. Yet, between October 1943 and mid-January 1944, the 99th had not encountered a single enemy aircraft!

That was about to change dramatically. The Squadron moved, with the 79th Group, to Capodichino Air Field near Naples on the west coast of "the Boot," on 16 January 1944, to cover the Allied landings at Anzio.

To break down German defenses in central Italy and to reach Rome fast, the Anglo-American VI Corps landed at Anzio on 22 January 1944. The enemy was caught by surprise and initially offered little resistance. The Mediterranean Allied Air Forces (MAAF) covered the landings and enabled Allied forces on the ground to dig in and consolidate their beachhead position. Unfortunately, the decision to dig in gave the Germans time to rush reinforcements to the area. Poor weather hampered Allied air interdiction of the Axis trains and truck convoys. A long and difficult campaign lay ahead.

For pilots of the 99th, the Anzio campaign provided opportunity to get back into air-to-air combat even while they were assigned close air support and interdiction missions. On 27 January, a flight of 16 of the Squadron's P-40Ls led by Lt. Clarence Jamison spotted a similar-sized flight of German FW-190s attacking Allied ships. The Warhawks dove on the enemy fighters, employing tactics learned from the pilots of the 79th, who were providing top cover. The 99th pilots stayed tight, holding their fire until the enemy was well within range of their six .50 caliber machine guns. Staying with their targets, they chased them 30 miles to Rome if necessary. The Axis fighters broke off their attack and fled; the 99th aircraft returned to Capodichino to amaze those on the ground with five victory rolls! Lieutenants Ashley, Baugh, Diez, Leon Roberts, and Toppins were credited with one kill each and Lt. Clarence Allen was officially credited with one-half. There were no losses to the 99th!

That same afternoon, three more Tuskegee Airmen entered the aerial victory column! Captain Custis and Lieutenants Bailey and Eagleson each downed a German FW-190 after a low-level dogfight and Major Roberts, the flight leader, knocked out a German machine gun position with the three machine guns he had left after being hit by flak. The day's jubilation over the eight aerial victories was not without sadness, however. Lieutenant Samuel Bruce became the first member of the 99th to lose his life in aerial combat.

Fierce fighting over Anzio continued the next day. The Squadron's victories continued, too. Diez and Lewis C. Smith each shot down a FW-190, Diez's second, and Charlie Hall increased his total of confirmed victories to three as he downed one FW-190 and one Bf-109. All of the 99th's P-40s returned safely to Capodichino; only one had suffered any damage. Four more victory rolls ended forever the doubts of the ground crews about the courage and skill of their pilots. They were beating a well-equipped, well-trained Luftwaffe in skies that had belonged to the Axis!

In the face of continued Allied ground attacks and under nearly continuous MAAF bombing and strafing, the Germans fought back—in the air and on the ground. Pilots noted increased enemy air activity and recognized quickly that they had to be more careful about chasing damaged German aircraft. On 5 February, two 99th aircraft were shot down; one pilot, Clarence Jamison, managed to bail out safely and to make it back to Capodichino. The other was lost. Lieutenant "Woody" Driver partially evened the score as he dove on a flight of FW-190s over Anzio beach, hit one on the right side of the cockpit and watched it crash as it struggled toward Rome. Two days later, the pilots of the

99th confirmed three more enemy aircraft without a loss. Eagleson scored his second and Leonard Jackson and Clinton Mills each added one. These were the last aerial victories for the P-40s of the 99th; it would be five long months before the 99th would register another air-to-air kill and then in a different aircraft.

While the excitement of the victories provided a temporary boost for the Squadron's morale, and Army Special Services was able to provide the men with additional recreational opportunities, the continuing winter weather made February and March bleak. When ceilings and visibility permitted, the 79th Fighter Group continued to pound away at enemy ground targets with their increasingly war weary *Warhawks*. Even when poor weather kept the planes on the ground, maintenance people continued 'round-the-clock efforts to keep the aircraft ready to fly, changing engines, patching holes, repairing damaged hydraulic systems and flight controls. Increasingly concerned that "the percentages" were catching up with them, the 99th's pilots took an especially active interest in the maintenance condition of their planes, checking them out even when

Pilots of the 332nd Fighter Group briefed before a combat mission. Weather, routes, targets, weapons, and communications procedures were reviewed in detail. *Official USAF Photo, U.S. Air Force Museum*

Mac Ross

99th Fighter Squadron pilot Mac Ross, one of the first group of black cadets to earn USAAF wings. *U.S. Air Force Museum*

When the 100th Fighter Squadron was the first to become active in the 332nd Fighter Group, it was headed by Lt. Mac Ross, a courageous pilot, who completed more than fifty combat missions during World War II.

A unique club was created—the Caterpillar Club—for all who used parachutes to save their lives and, when Ross bailed out of his P-40 at TAAF, he became the first black member. Ross, a graduate of West Virginia State College, earned his private pilot's certificate through the college's civilian pilot training program and enlisted as an Aviation Cadet in 1941.

Harold Sawyer said of Mac, "I would say he was the type of person who never wanted to fail at anything. When we arrived overseas, Mac had been flying the twin engined UC-78 transport to take Colonel Davis back and forth to headquarters. He was a good pilot, but when in the process of transitioning into the P-51—I don't know what happened, whether the engine went dead on him or what—but he crashed." Captain Mac Ross was killed in that fiery crash, graphically demonstrating the dangers of war for which Ross and the others "fought for the right to fight."

Mac Ross received the Distinguished Flying Cross, the Legion of Merit, numerous campaign medals and the Purple Heart. Posthumously, the U.S. Postal Service, Dayton, Ohio, dedicated the Mac Ross Memorial Philatelic Room, 27 June 1989. The published tribute said, in part:

"No greater can a man be than to give up his life for his country. There's an idealism there; a camaraderie, a brotherhood, for all who served their country. A young black man, one of ten children, a college graduate, becoming one of the first black pilots in the U.S. Air Force, then a Commander of a Pursuit Squadron, manifests these three hallowed words echoed in time; obedience, integrity and courage. As a role model for today's youth, Mac Ross has stood the lasting test. The indomitable and unflagging determination of Captain Mac Ross in 1944 is still applicable in 1989 and will be in the year 2000. It is men like him who are the bricks and stones that make our country great. …Mac Ross is a symbol: of what men can achieve and what they should strive for. The men and women of Tuskegee have gone down in history as true heroes of World War II. From the original five, Mac Ross, Benjamin O. Davis, Jr., Charles DeBow, George "Spanky" Roberts and Lemuel Custis, grew a strong "air force" numbering 992 graduates of the Tuskegee Flying School. They served this country well and proved what could be achieved against all odds.

A salute to the Tuskegee Airmen."

they weren't being inspected and developing even closer personal relationships with their crew chiefs.

The coincident Luftwaffe raid on the Capodichino Air Field and the eruption of Mount Vesuvius on 15 March 1944 caused havoc in many quarters, but the 99th came through both unscathed. They continued their close air support and interdiction missions against the dogged Germans.

April brought more major changes for the Squadron. After six months of very valuable association with Colonel Bates and the 79th Fighter Group, the 99th was re-assigned first to the 324th and then, briefly, to the 86th Fighter Groups. Neither affiliation was as pleasant or as productive as with the 79th. Major Spanky Roberts completed 78 combat missions with the Squadron, many as Operations Officer and even more as Squadron Commander, and was returned to the United States for a well-earned rest. He was replaced by Capt. Erwin Lawrence, member of Class 42-F at Tuskegee and an experienced and capable fighter pilot. Coincidentally, the 99th moved again, first to Cercola Field on 2 April and later to Orbetello.

With all the changes, the missions remained the same—close support of the slowly advancing ground forces and bombing raids on the enemy's supply routes. Even in the thick of furious fighting around the town of Cassino, no enemy aircraft were engaged, and the 99th had to be content with its role in the Anzio campaign, the Battle for Cassino, and the fall of Rome on June 4th. In stark contrast with reports filed at the end of the Pantelleria and Sicily campaigns, this time the Squadron could take great pride in the praise of its senior officers. On 20 April, Gen. Ira Eaker, Commander of the Mediterranean Allied Air Force, commended the 99th in these words: "By the magnificent showing your fliers have made since coming to this theater, and especially in the Anzio beachhead operations, you have not only won the plaudits of the Air Force, but have earned the opportunity to apply your talent to much more advanced work than was at one time planned for you."

The Tuskegee Experiment's pay-off was becoming more widely known and appreciated. In their first year in combat, the men of the 99th Fighter Squadron flew 500 missions made up of 3,277 sorties. They downed 17 enemy aircraft and damaged many more in the air. They lost eight pilots—two in air-to-air engagements and six more in ground attack missions. They provided close air support in five major campaigns and inhibited the enemy's ability to reinforce, re-equip, and resupply critical ground forces. Significantly, they successfully withstood the verbal sneak attacks of their senior Army Air Force leadership. In the "separate but equal" air force, it was time to join their compatriots at Ramitelli and to get on with winning the war.

332nd Fighter Group, October 1942–November 1943

The 332nd Fighter Group was activated at Tuskegee Army Air Field on 13 October 1942. To be made up of a Headquarters staff, the 332nd Fighter Control Squadron and the 100th, 301st, and 302nd Fighter Squadrons, the new Group got off to a slow start. At the end of October, there were only two officers and seven enlisted men assigned or attached. A month later, the numbers had grown only slightly. White officers and enlisted personnel provided an initial cadre in an "attached" status and qualified blacks were "assigned" as they became available from other organizations, some of which were being disbanded, and from schools across the country.

On 3 December, Maj. Sam W. Westbrook, Jr.'s acting commander status became permanent; it fell to him to answer the pressing question: Where were all the people to fill out four squadrons and a headquarters going to come from?

The first, partial, answer came in January 1943, when 19 brand new 2nd Lieutenants arrived directly from the integrated Officer Candidate School at Miami

Charles P. Bailey, 99th Fighter Squadron. Note the "A" (designating the 99th Fighter Squadron) on the P-40 in the background. *U.S. Air Force Museum.*

Beach, Florida, and, from then on there was a steady stream of new personnel—officers, non-commissioned officers and enlisted men. At Tuskegee, there was great pressure to "join up" with the new organization even before the March 1943 announcement of the Group's move to Michigan. When that move actually began, the Group's total personnel had reached 75 officers and 934 enlisted men and two troop trains were required to transport all but the ten pilots who flew their P-40s north.

Intensive training for combat began in earnest in April 1943, under the supervision of the all-white 403rd Fighter Squadron. The leaders of Detroit's black community came out to Selfridge Army Air Field to welcome the Group and to see, first-hand, what that training was all about. By the end of the month, the Group's composition and personnel status was:

Enlisted

Organization	Commander	Officers	Personnel
332nd Headquarters	Lt. Col. Westbrook	15	33
100th Fighter Sqdn	1st Lt. George Knox	34	253
301st Fighter Sqdn	1st Lt. Charles DeBow	13	250
302nd Fighter Sqdn	2nd Lt. Wm. Mattison	6	291

April saw a steady stream of new arrivals including newly commissioned 2nd Lt. James Pughsley, a "Buffalo Soldier" who served his country in World War I and who had been stationed in Arizona, Texas, and Oklahoma between the wars. First Lieutenant Knox replaced Mac Ross as Commander of the 100th and the first move, to Iosco County's Oscoda Army Air Field on Lake Huron, took place in spite of a protest from the Iosco County Board of Supervisors.

The Group Historian reported in May 1943: "Over the sandy pine shores of Lake Huron, pilots of the 332nd Fighter Group polished up their flight and gunnery tactics for that inevitable test of their skill and courage which will eventually lead them to

Key players in the combat effectiveness of the unit, aircraft mechanics labored to keep the fighters ready to fly and fight. *Official USAF Photo, U.S. Air Force Museum*

General Daniel "Chappie" James

For those in the audience whenever history's first black four-star Gen. Daniel "Chappie" James spoke, the man seemed a giant. His towering presence and his commanding voice would have been sufficient to rivet the attention of his listener. But, it was his eloquent words, many of them attributable directly to one of the greatest influences of his life—his mother—that captured even the most reticent. Mrs. Lillie A. James, mother of seventeen children, the youngest of whom was Chappie, must have been a forceful parent. She started her own school when she felt that her children and others were not getting a good education, leading her children and others on a firm and deliberate path toward achievement and excellence. She taught, "Thou shall not quit," and "Don't be so busy practicing your right to dissent that you forget your responsibility to contribute." When her youngest son spoke those words so eloquently time and time again, those who heard could be likened to the widening circles that emanate from a pebble dropped into still water. There is no way to measure the extent of the influence of this valiant lady and her prominent, highly successful son.

James attended Tuskegee Institute as a student, majoring in physical education and obtaining his private pilot's certificate in 1942 through the civilian pilot training program. The same year, James married the former Dorothy Watkins and served as a civilian flight instructor in the Army Air Corps Aviation Cadet Program at Tuskegee Army Air Field. In 1943, he became a cadet in the program, graduating in July and pinning on the wings of the U.S. Army Air Corps as well as the insignia of a 2nd Lieutenant.

Moved to Selfridge Field with the 477th Bomb Group, James trained in medium range, multi-engine B-25 bombers. With Truman's integration order in 1948, he was assigned to command the 12th Fighter Bomber Squadron at Clark Field, Philippine Islands. James flew 101 combat missions in P-51 and F-80 aircraft in Korea between 1950 and 1951. He attended Air Command and Staff College at Maxwell Air Force Base, Montgomery, Alabama, which was followed by an assignment to USAF Headquarters, Washington, DC in the Air Defense Division.

Assigned for a decade with tactical fighter units, James moved from the United States, England, and Libya to the Vietnam War. He flew 78 missions in F-4C fighters from a base in Thailand, followed, in 1970, by assignment to the Office of the Secretary of Defense as

Deputy Assistant Secretary. It was in this capacity that the eloquent orator created a lasting reputation as a public speaker.

From Pensacola, Florida, 1920, where the young Chappie was born; from the values and goals instilled by his mother, who insisted, "Don't ever turn your back on your God, your country, or your flag": from Tuskegee to the Pentagon, James went from aviation cadet to history books. He advanced from Brigadier General to Major General, from Lieutenant General and finally to pin on the coveted fourth star. General James retired from the last assignment of his career as the Commander in Chief of North American Air Defense and Aerospace Commands. His death within one month of retirement from the U.S. Air Force ended a meteoric and proud career.

Tuskegee Airmen working on the engine of a P-40. Keeping them flying meant working long days and nights in all weather conditions. *Bernard S. Proctor, U.S. Air Force Museum*

pursuit and battle in the skies of Europe and Asia." Sadly, the polishing of those tactics led to the untimely deaths of two of the 100th Fighter Squadron's young pilots, Lieutenants Jerome Edwards and Wilmeth Sidat-Singh; stark reminders of the hazards to be encountered in preparing for a challenging future.

Colonel Robert R. Selway, Jr., a West Point graduate in the Class of 1924 and a senior pilot, assumed command of the Group on 16 May; Lieutenant Colonel Westbrook became his Executive Officer. They would oversee the Group's preparations for combat. Among their first tasks was the move of the 301st to Oscoda, to follow the 100th onto aerial gunnery and bombing ranges, and the establishment of an Air Intelligence School at Selfridge. They also had the pleasure of seeing their troops feted by Michigan's most attractive young ladies at parties.

Colonel Selway's training began to take effect in June. A "rainy day" program was established to ensure training continuity and a number of new Boards of Officers were established. Two of those Boards—a Flying Evaluation Board and an

Instrument Flying Proficiency Board—were set up in reaction to the worrisome number of accidents; two more 100th Fighter Squadron officers, Lieutenants Hill, a pilot, and Blakeney, the weather officer, were killed when Hill became lost in a rapidly forming fog and crashed into the lake.

The steady influx of officer and enlisted personnel continued with pilots and ground officers arriving from Tuskegee; mechanics from Lincoln, Nebraska, Buffalo, New York, and Chanute Field, Illinois; armorers from Denver, Colorado; radar mechanics from Tomah, Wisconsin; and communications personnel from Fort Monmouth, New Jersey, and Camp Crowder, Missouri. Over 500 people were transferred in and out of the Group and the 403rd, supervising the training program, was severely tested, even after the creation of the 332nd Provisional Squadron to which unassigned personnel were directed. As the Group historian put it: "What was once an organization in dire need of personnel became a replacement center for stations throughout the country."

Compounding the challenges, the Group's P-40Cs and Fs were being replaced by P-39

Airacobras. The retirement of the P-40s was a relief to the maintenance people for, as Master Sgt. James Jones, maintenance line chief of the 100th, said, "It was all my men could do to keep them airworthy. We lived in continual fear that someone wouldn't return due to a failure beyond our control. When they returned from a flight, it appeared frequently as though they'd flown through an oil storm."

By early September, the conversion was complete. Although pilots were pleased to have something newer than early-vintage P-40s to take to the Oscoda Bomb and Gunnery Range, they were disappointed that the persistent rumors of P-47 *Thunderbolts* hadn't come true. A rumor that did come true was the 7 October 1943 appointment of Lt. Col. Benjamin O. Davis, Jr., as Commander of the 332nd Fighter Group replacing Colonel Selway.

Predicted by "Spratmo" (a system of gathering information that is unparalleled), the 332nd historian described the impact of Colonel Davis' appointment in his October report: "With a broad smile on their faces, it was indicative that each individual of the 332nd was proud over the appointment of the new commandant. Having anticipated his coming quite keenly, his arrival gave the organization a tremendous lift."

Colonel Davis assumed command at a critical time. The influx of new people and new aircraft strained the facilities at Selfridge and Oscoda and flying from dawn to dusk to meet combat readiness training requirements resulted in some hazardous and some humorous incidents. One daring exploit belonged to Lt. Spurgeon Ellington of the 100th when he became separated from his formation. Descending through an overcast, he spotted a familiar four-lane highway and decided that, since he was low on fuel, he'd make a forced landing. Circling to find a break in traffic, he picked his spot, made a perfect landing, and boldly taxied his P-39 to a gas station to "fill it up."

To add a little to the pressure already being felt by the Group, warning orders were received on 19 October 1943 stating that the 332nd would be prepared to proceed to the Port of Embarkation at the call of the Port Commander. It had now become a question of *when* not *whether* and final preparations were hurried.

War on the Home Front

As the 99th Fighter Squadron continued the fight against fascism in the skies over Italy, and the 332nd accelerated its preparations to join that fight, the war against segregation heated up at home. Black Americans took up a "Double V" battle cry, as determined to achieve victory over *Jim Crow* as over U.S. enemies abroad; they focused their anger and their frustration on the war effort and on the difficulties they experienced in becoming full participants. A common wartime graffiti was "Here lies a black man, killed fighting a yellow man for the glory of the white man."

The Tuskegee Airmen fought for the right to fight America's enemies abroad. Other blacks fought against segregation on the streets of their hometowns—in Harlem and Detroit, in Houston, Charleston, Richmond, Philadelphia, and in the nation's capitol—and at stateside military bases—Fort Jackson, Fort Bragg, and Fort Dix.

Where ugly race riots had historically resulted from white reactions to accusations that a black had committed a crime against a white, wartime trouble was generally initiated by blacks in increasingly violent protests against the limits that segregation imposed upon them. Incidents involved groups of civilians as well as military people; both were angry and frustrated that their participation in the war effort was restricted and that *Jim Crow* was having an even greater impact in their lives. They were angry that their hopes for a better America for all races were thwarted by the intransigent attitudes they encountered. The hope that fuller participation

Mac Ross, transitioning to a new aircraft in Italy, lost his life in this tragic crash of a P-51 on 11 July 1944. *Lt. Col. Alexander Jefferson, USAF Retired, Tuskegee Airmen National Museum, Detroit, MI*

in the life of their homeland would be a result of the war seemed supported by the Roosevelt administration, shared by leaders of the black community, and widely reported in the black press with its circulation of two million.

That hope was clearly *not* supported by many of America's wartime leaders including those who reluctantly supported the inception of the Tuskegee Experiment, who supervised the training of black airmen and technicians, and who employed their combat capabilities. Luckily, there were exceptions.

One was Maj. Gen. George E. Stratemeyer, the chief of the Air Staff. Although statements attributed to him in the early discussions of training for black airmen would indicate a racial attitude much like the majority of his peers, he had clearly changed his mind by the time he came to his key post in the Pentagon. Even before Judge Hastie's resignation as civilian aid on Negro affairs, often cited as the principal catalyst for a significant change in racial policies of the War Department, General Stratemeyer had assumed personal control of the Army Air Force's race relations policies and programs. Having requested the gathering of meaningful data on the treatment and training of Negroes in the AAF, he not only ordered a halt to the plan to develop a segregated Officer Candidate School at Jefferson Barracks, Missouri, he ordered an end to all segregated training except that conducted at Tuskegee. His actions reflected a major change in thinking in the War Department and among the Army's senior leadership.

No longer was the black American thought to be biologically inferior; rather, he was seen as the victim of economic, educational, and social deprivation flowing from prejudice and, even, institutionalized racism. This was an enormous change, one that would require major cultural and educational efforts which the War Department was prepared to begin. Pamphlets like *Command of Negro Troops*, manuals like *Leadership of the Negro Soldier*, and films like *Wings for This Man*, narrated, Stanley Sandler tells us, "by the well-known Hollywood liberal, Capt. Ronald Reagan, USAAF," were prepared and widely distributed. Broad training programs, based on these materials, were conducted. Integrated training for black flight surgeons and for black navigator and bombardier trainees was initiated. The Officer Candidate School at Miami Beach, Florida, continued to operate on a largely integrated basis in spite

Bernard S. Proctor in the lead at an overseas track meet. Sports provided a welcome break from combat and a positive morale factor. *Bernard S. Proctor, U.S. Air Force Museum*

of local laws which forbade the presence of blacks. Exceptions to full integration included: sleeping quarters were segregated, black students were not permitted to perform in roles in which they "commanded" whites (not including the few black instructors), and black students had to be driven to Miami for weekly haircuts since there were no barbers in Miami Beach who would cut a black man's hair. Informal histories document the case of the officer candidate, later an Air Force Major General, who was taken to Miami in a staff car to get his hair trimmed. One can only imagine how he felt to be riding alone in the back seat.

It must be noted that, even as action signalled attitudinal change at the policy-making level and as military leaders promulgated programs and training to support their new view, effecting change at the "grass roots" level would take generations. Segregation con-

tinued to impact black soldiers and airmen wherever they went in the United States for many years, more in some places than in others. No black serviceman or woman escaped it; none will ever forget it. One Tuskegee Airman said that when they get together, they talk about how they were treated—not about their airplanes or their combat experiences. It is a tribute, albeit an insufficient one, to their courage and determination to fight for their homeland with all its flaws that the "grass roots" changes have taken place at all. These valiant aviators established a valuable pattern.

A second exception was Colonel, later Brigadier General, Noel Parrish, the Tuskegee Commander. On 1 April 1942, a black MP from TAAF took a black prisoner at gunpoint from a white Tuskegee town policeman. The town police reinforced by sheriff's people, state police and a gang of armed local white citizens, fought back, disarming the

Lemuel R. Custis, Wilson V. Eagleson, Willie Ashley, and Charles B. Hall of the 99th Fighter Squadron. Lieutenant Hall's was the first aerial victory for the 99th. He enjoyed a precious ice-cold Coca-Cola, the prize that had been specifically set aside for the first victor. *Bernard S. Proctor, U.S. Air Force Museum*

Crew members in Sicily supported both General Patton's U.S. Seventh Army and General Montgomery's British Eighth Army. *Bernard S. Proctor, U.S. Air Force Museum*

Military Police patrol, badly beating at least one of the MPs, and retrieving the prisoner. Led by Parrish, white soldiers from TAAF went into town and succeeded in gathering most of the black soldiers together, returning them to the base. Parrish's courage and leadership defused a potentially disastrous situation. Army investigators later concluded the black MP had been wrong in taking the prisoner forcibly and that the police had used excessive force. Had this crisis reached the proportions of the riot incidents at Fort Jackson or Fort Dix, there might have been an untimely end to the entire Tuskegee Experiment.

That same courage and leadership ability was repeated as Parrish fought with the Pentagon for what he knew to be right, as he worked with local officials to maintain a peaceful relationship between the town and the base, and as he dealt with the overcrowding, confusion, and poor morale that stemmed from Tuskegee's status as the only training base for blacks in several military career fields. Parrish, in turn, credited Tuskegee's unique ability to focus on its military aviation training mission to instructors and students: their serious effort and abilities to escape minor distractions. The clear focus on the job to be done characterized Colonel Parrish's style and supported his effectiveness. That he was able, for so long and in the face of so much opposition, to develop that same focus in his staff, the instructors, and the myriad students who came under his command is the mark of a true leader.

War against *Jim Crow* at home, which had little impact until some time after the war against fascism was over, took many forms. The violence, rhetoric, and leadership of men like Stratemeyer and Parrish all played a role in changing entrenched prejudices and institutional racism that resembled that in fascist nations against which the war abroad was fought. No factor was greater, however, than the determination to succeed and to contribute demonstrated by the men and women of Tuskegee.

332nd Fighter Group, December 1943–June 1944

A movement order, dated 15 December 1943, directed that the 332nd Fighter Group Headquarters and the 100th, 301st, and 302nd Fighter Squadrons "will move at authorized strength via rail from Selfridge Field to Camp Patrick Henry, VA, so as to arrive thereat during daylight hours of 24 Dec 1943." Last minute training was accomplished under the eyes of inspectors. Records were reviewed and brought up to date. Hangars, normally devoted to aircraft maintenance, converted to warehouses in which the myriad items of organizational equipment were wrapped and packed in waterproof containers for shipment overseas. Pre-Christmas, pre-Departure parties were joyous affairs! The ladies wore their most colorful outfits. Farewells were tearful even as an air of excitement pervaded the crowds. With winter winds coming off Lake St. Clair and temperatures hovering near zero and with the Post Band providing lively marches as background, the well-trained—some said overtrained—Tuskegee Airmen departed. The first train, carrying the Headquarters and the 100th Fighter Squadron and commanded by Lt. Col. B. O. Davis, Jr., arrived at Camp Patrick Henry just before noon on Christmas Eve. The second, under Capt. Charles DeBow and with the 301st and 302nd Squadrons aboard, pulled in, 23 hours later, on Christmas morning.

Joy of the season and the excitement of finally being on their way to combat were dimmed for the men of the 332nd by one last insulting incident. Going to the Base theater, the airmen found that a section had been roped off for blacks. A developing incident was squelched by Colonel Davis' quick action in confining his troops to quarters. He then went to the Base Commander and warned that he could not be responsible for any further actions by his people if segregation at the theater wasn't ended. The ropes came down at the movies, but other Base facilities remained segregated or closed to the men of the 332nd. Captain Vance Marchbanks, the Group Surgeon, noted that, in contrast to celebrations for their white brothers-in-arms: "The wounds to our patriotic pride would take a long time to heal and, when they did heal, they left a deep and ugly scar. About the best thing that could happen would be an early departure."

After last hour telephone calls to wives and mothers and sweethearts, for which they had to wait in line for up to six hours, the 332nd was ordered to be ready—with full field pack, shelter halves, and blankets in a horseshoe roll—to clear the staging area before dawn on the 3rd of January, 1944. The Group Historian wrote: "It may be a meteorological phenom-

Major George S. "Spanky" Roberts who assumed command of the 99th Fighter Squadron, replacing Lieutenant Colonel Davis, in August 1943. Note the patch of the 99th Fighter Squadron. *U.S. Air Force Museum*

enon or just pure coincidence but with each move that I have witnessed for the 332nd, there has always been a forerunner, rain, rain, and more rain. At approximately 0445, 3 January 1944, the 332nd Fighter Group was at the Rendezvous Point amid an intermittent sprinkle of raindrops which in a very short while developed into an incessant downpour which lasted until—well to be exact, all day. After two and one half hours of standing with full field pack, duffle bag, and all its accessories, the Group finally moved off, soaked from head to foot with not a dry stitch of clothing anywhere. A brief ride on the *Chattanooga Train* ended with the Group at the docks of Hampton Roads, Va. An arousing welcome was accorded the men by the Port Band playing the Air Corps song and other hot tunes. In a short while all personnel were aboard the ferry boat *Mohawk* on their way to those dear old 'Liberty Ships' lying across the bay. After the usual procedure of waiting, the Headquarters Squadron along with units of the 96th Service Group boarded the *SS William Few —The Billy Foo* as the ship was later rechristened by members of the 332nd."

Colonel Bates, C.O. of the 79th Fighter Group shown with Captains Spanky Roberts and Charles Hall, Lieutenant Eagleson, Capt. Lemuel Custis, and Maj. Gen. John Cannon. The 99th Fighter Squadron's assignment with Colonel Bates' 79th was a time of great learning and great success. *Bernard S. Proctor, U.S. Air Force Museum*

Captain Howard Baugh, Maj. George "Spanky" Roberts, and Capt. Erwin Lawrence of the 99th Fighter Squadron. A veteran pilot with the 99th, Captain Lawrence lost his life on a strafing attack on a German airfield. *U.S. Air Force Museum*

The crossing was described as uneventful, with only a submarine scare to break the monotony of the month-long voyage through rough choppy waters of the North Atlantic. The men peeked at the lights of Gibraltar as the ships passed through the Straits and land wasn't sighted again until they were approaching Augusta, Sicily, on the morning of 27 January 1944. The Historian closes out his January report with the notation that the *Billy Foo* is destined for Taranto, Italy. The Group Headquarters Morning Report for the end of the month included three civilians, a War Correspondent for the *Norfolk Journal Guide*, a correspondent and cartoonist for the *Pittsburgh Courier,* and a representative of the *American Red Cross*.

Finally, the SS *William Few* reached the harbor at Taranto on 3 February and the men of 332nd Headquarters Squadron set foot on dry land. They moved off to the staging area five miles away on their first hike in some time; their aching feet relieved by warm welcomes of the men of the 301st and 302nd Squadrons. The question was: where was the 100th FS? Had they fallen victim to Axis submarines or to Jerries' aerial raiders? The answer (and the reunion!) finally occurred on 8 February when, after a 13 1/2 hour motorized convoy trip, the 332nd, 301st, and 302nd reached Montecorvino Air Base and discovered that the 100th was not only present for duty, it had been engaged in combat missions since the 5th of the month!

The airmen also discovered that the optimistic rumors of aircraft transition to the P-47 *Thunderbolt* or P-63 *Kingcobra* or P-38 *Lightning* hadn't come true. The 332nd was going to keep flying the P-39Q on "convoy escort, harbor protection, scrambles, and point patrol extending from Cape Palenuro and the Gulf of Policastro to the Ponziane Islands" under the 62nd Fighter Wing, 12th Air Force. They replaced the 81st Fighter Group which transferred out to move to the China-Burma-India Theater as part of the 14th Air Force under Brig. Gen. Claire Chennault.

By 17 February, all three Squadrons were operational; they flew over 200 combat sorties totalling 236 operational hours in flight. The 301st experienced its first loss when 2nd Lt. Harry J. Daniels failed to return from a mission flown in bad weather, and the 302nd had its first contact with the enemy when, on the 15th and again on the 17th, the pilots of the Squadron learned to their chagrin that the P-39 just didn't have the power to close in on the German Ju-88s they were chasing. On both occasions, they scored hits on

Private John T. Fields, 100th Fighter Squadron, arms a North American P-51 Mustang with .50 calibre machine gun ammunition. *U.S. Air Force Museum*

the enemy airplanes, but were unable to finish the job. On 21 February, the 100th was relocated to Capodichino Airport to continue its operations and Group Headquarters and the 302nd followed in March and in April.

As experience grew, the Group expanded its missions and its capability and higher headquarters planned to re-equip it with the P-63, the updated and at least theoretically more capable version of the *Airacobra*. But delays in production of the new airplane and General Eaker's revisions to the plans resulted in the provision to the 332nd of razorback P-47Ds, the first six of which arrived on 25 April. They were being transferred to the Group from the 325th which was transitioning to P-51 *Mustangs*, WWII's premier reciprocating engine fighter. Their black and yellow checkerboard tails were hastily overpainted with the 332nd Fighter Group's newly designated all red paint and pilots began their training in the "Jug," as the P-47s with their big radial engines were known.

May brought a number of major changes to the 332nd Fighter Group and its three Squadrons. Not only were the pilots learning to fly, and the mechanics learning to service, a new aircraft painted with a new and distinctive paint scheme, but major organization, location, and mission changes developed. On 22 May, the Group was re-assigned from the 62nd Fighter Wing, 12th Air Force to the 306th Fighter Wing, 15th Air Force.

Crew Chief, S/Sgt. William Accoo, washing prior to waxing. Note his reflection in the clean surface of "his" airplane! *U.S. Air Force Museum*

Lieutenant Clarence Jamison, 99th Fighter Squadron flew, trained, and fought with the 79th Fighter Group while stationed at Foggia, Italy. *U.S. Air Force Museum*

lap, and having said goodbye to old Vesuvius, we journeyed by truck convoy to Ramitelli Air Field to become a part of the great Fifteenth Air Force. The air was thick with excitement, with conjecture, with rumor, with lies about future activity in our new Air Force. It was also filled with the hope of all that the opportunity had at last presented itself to gain those successes which are so cherished by flying men. The hot sun blazed down on Ramitelli, the parched earth, the dust, and the great distance that separated us from Naples, our former playground, in no way affected our newly found spirit. We kept saying 'Goodbye Harbor Patrol, Goodbye Convoy Patrol, and Welcome *Thunderbolts.*'"

Flying the *Thunderbolt* brought major changes of its own. In his journal, Gene Carter, pilot and Maintenance Officer of the 99th, wrote: "We are changing over to P-47s now. You should see me. I look like the man in the whale. I can nearly stand up and walk around in the cockpit, but I like the big heavy thing."

Harold Sawyer, 301st Fighter Squadron pilot, added this description of one P-47 operation: "Every time you took off in the P-47, you took off over the Adriatic Sea. The Italians drove carts and things and, one day, a guy with a horse and cart was going across the end of the runway. Those P-47s had auxiliary tanks on the wings, we were loaded down. I mean, a P-47 is a heavy airplane anyway and with those aux tanks on there, man, you need all the runway that you can get. Most of the time, you were barely gettin' off the ground by the end of the runway. This guy was going across there. I don't know whether the pilot knew whether he was clearin' him or not, but he went in and that P-47 exploded with all that black smoke. Once you start, one right after another, you don't miss a beat. Once they take that flare gun and shoot it up in the air, you're gone!

With the change came its new role as escort for long-range, heavy bombers as they carried out critically important strategic attacks on supply centers, transportation hubs, and harbor installations. On the 26th, Group Headquarters and the 301st and 302nd moved to Ramitelli Air Field, where they were joined by the 100th on the 9th of June and by the re-assigned 99th Fighter Squadron on 6 July.

June's Group Historian's report began: "After having dumped our *Airacobras* into someone else's

Pilots Howard Baugh, Jack Rogers, John Gibson, Herky Perry, and Willie Fuller debriefing after a combat mission. *Lt. Col. Gene Carter, USAF Retired*

You're just constantly rolling. With all that smoke and everything, you didn't know what was ahead of you. I never will forget that. But, you couldn't stop and think about it. You were on your way. After you got up into the air, you just went on."

The *Red Tails* flew their first combat mission in the P-47 on 7 June. Two days later they had their first really big day, a day filled with the cherished success anticipated. The 302nd's Lt. Wendell O. Pruitt, credited with the first kill that day, described it: "We were assigned to fly top cover for heavy bombers. On approaching the Udine area, a flock of Bf-109s were observed making attacks from 5 o'clock on a formation of B-24s. Each enemy aircraft made a pass at the bombers and fell into a left rolling turn. I rolled over, shoved everything forward, and closed in on a -109 at about 475 mph. I waited as he shallowed out of a turn, gave him a couple of two second bursts, and watched him explode."

Aside from Pruitt's kill, the 301st's Frederick Funderburg was credited with downing two Messerschmitts, and Lieutenants Melvin Jackson and Charles Bussey, both of the 302nd, were each credited with one. Pruitt's crew chief, staff Sgt. Samuel Jacobs, related what happened when A Flight returned to Ramitelli. Some representatives of Republic Aviation, the P-47's manufacturer, and some Air Corps brass had come to the field to teach the 332nd to fly and crew the *Thunderbolt.* Sergeant Jacobs recalled: "...this Major was standing atop a munitions carrier telling us 'boys' all about the 'flying bathtub' and how it should never be slow rolled below a thousand feet, due to its excessive weight. No sooner had he finished his statement than 'A' Flight was returning from its victorious mission. Down on the deck, props cutting grass, came Lieutenant Pruitt and his wingman Lee Archer, nearly touching wings. Lieutenant Pruitt pulled up into the prettiest victory roll you'd ever see, with Archer right in his pocket, as the Major screamed, 'You can't do that!'"

Fifteen days later the Group, and the 100th Fighter Squadron, sustained a very painful loss. Captain Robert Tresville, a member of West Point

William A. "Wild Bill" and Steve Campbell, Father and Son

Colonel William A. Campbell, one of the first black airmen to be honored with the award of two Distinguished Flying Crosses, became commanding officer of the squadron he helped to make famous—the 99th. During the invasion of the island of Pantelleria, Campbell was credited as the "first" black to wage aerial combat for U.S. Army Air Forces. He said, "I simply flew on the wing of a white pilot. I was told, 'When I dive, you dive. When I release my bomb, you release yours.' I may have received the notoriety of being the first, but it was only by a matter of seconds."

Campbell participated in Sicilian and Italian campaigns, then returned to Selfridge Field, Michigan, to the 553rd Combat Crew Training Squadron. He rejoined the 99th at Ramitelli, Italy, in August 1944 and transitioned to the P-51 Mustang. During his second combat tour, Campbell commanded the squadron. Credited with 106 combat missions (36 in P-40s and 70 in P-51s), two DFCs and ten Air Medals, Campbell, who also downed a Bf-109 in aerial combat over Austria, said, "Fighting in an airplane is different than face-to-face. You felt elated when you got a kill."

He and eleven other pilots were airborne when enemy aircraft appeared. "We chased them, very close to the ground. We had to be careful. They must have been new pilots because one of them pulled up and that was the wrong thing to do. I closed on him immediately and, while he was climbing, I fired. He nosed over, dove straight down, and crashed."

Campbell graduated from Tuskegee Institute in 1937 and his is a proud legacy. A bust of his father, Thomas Monroe Campbell, was sculpted by Isaac Hathaway and is displayed at the George Washington Carver Museum, Tuskegee. Presented by fellow extension agents, the sculpture honors Thomas Campbell as "first Negro Extension Agent appointed by the Federal Government." Thomas Campbell served farmers of his region from 1906 to 1947.

Sharing his legacy was his sister—Captain Abbie Noel Campbell, Executive Officer, 6888 Postal Battalion, the only Negro WAC Battalion to serve overseas in World War II—and his brother, Captain Thomas M. Campbell, who served in France and Germany during 1944 and 1945 as Battalion Surgeon, 614th Tank Destroyer Battalion.

Bill Campbell and his wife, Wilma, have three sons: Willam A., Jr, a lawyer in Atlanta; David B., assistant basketball coach at Kansas State; and Stephen C., a former F-15 pilot and current captain with Southwest Airlines with whom he is pictured in La Grone's painting. La Grone said, "Before Steve Campbell got out of the Air Force, he arranged a flight at Luke Air Force Base for his father. Because they were father and son, they were not allowed to fly in the same aircraft. The squadron commander piloted the F-15 that carried Bill in the back seat and they flew close formation with Steve piloting the lead F-15. It was a proud and touching moment that I wanted to capture."

Post-World War II, in August 1947, Campbell became Group Commander and later led a team from his unit to first place in the conventional fighter division of the first worldwide gunnery meet. After earning a Master's degree in June 1950, he was assigned to Headquarters, Far East Air Force, Japan.

Completing Air Command and Staff College at Maxwell Air Force Base, Alabama, he checked out in the North American F-86 Sabre and reported to the 51st Fighter-Interceptor Wing, Suwon, Korea.

In March 1954, he became Commander of the 25th Fighter Interceptor Squadron and moved with the squadron to Okinawa where he stayed until February 1955. He left to serve as Professor of Air Science at Tuskegee Institute.

Assigned in Turkey and at the Pentagon, he became Director of Operations, USAF Advisory Group, Military Assistance Command Vietnam, Tan Son Nhut AB. He flew 40 missions in Douglas A-1 Skyraiders of the Vietnamese Air Force. He served at Ent Air Force Base, Colorado, where he retired as a Colonel, USAF, in 1970. During his career he amassed nearly 6,000 hours in 16 different aircraft. Settling in California, he became a Professor of Management at the Naval Postgraduate School, Monterey, CA.

One of the most admired Tuskegee Airmen, Bill Campbell exemplifies the cool competence of a successful pilot and man. He said, of the indignities and bigotry faced at every turn, "I don't feel any bitterness about that time period. I was young. I had a good time being a pilot. I liked what I was doing."

When honored as an Eagle at Maxwell Air Force Base's annual Gathering of Eagles, 1994, he displayed his generous sense of humor. With a smile, Campbell said, "If we had known the significance of our actions in World War II, we would have paid more attention to what was going on." Then he added, "Not many Lieutenants worry too much about strategy."

As advice for others, Campbell said, "Do the best that you are capable of doing. That's all you can do. But, do it all the time."

Class of 1943 and a recent Tuskegee AAF pilot training graduate, who had just assumed command of the 100th, was lost in the Tyrrhenian Sea on a low, level mission of his own design. The flight's objective was to stay low enough to avoid enemy radar detection en route to a strafing attack against critical enemy supply lines. The four plane flight stayed "on the deck" the whole way; approaching Corsica, the ceiling fell in and the flight had great difficulty maintaining formation. As they approached the northern Italian coast, Captain Tresville apparently became momentarily disoriented. The moment was all it took! He crashed into the sea. The strafing attack never took place and a valued member of the Tuskegee Airmen lost his life.

The Tuskegee Airmen's last P-47 achievement was unique! It happened on 25 June when a flight of five, led by Captain Joseph Elsberry, was returning from a strafing mission flown against the Istrian Peninsula in Northern Italy. They spotted a German destroyer in Trieste Harbor and dove on it. Lieutenant Pruitt made the first hit, which set the warship afire, and Lt. Pierson apparently made a direct hit on the ship's ammunition storage area, for the destroyer exploded and started for the bottom almost immediately. Leaders at 15th Air Force were skeptical that fighters, without bombs, could kill a destroyer, but the gun camera evidence proved that the 332nd had done what no other Group had ever accomplished!

As the Group was growing accustomed to the P-47, another aircraft transition was begun and red paint was brought out for a second time. The 31st and 325th Fighter Groups transitioned to the P-51D and the 332nd received their P-51Bs and Cs, repainting them with their distinguishing red tails and with special Squadron markings on trim tabs and nose bands, just aft of the solid red spinners. When the 99th joined them in July, the 332nd became the only Group with four squadrons and only one of two in the Army Air Corps to be totally segregated.

477th Bombardment Group

According to official lineages of World War II organizations, the 477th Bombardment Group (Medium) was constituted on 13 May 1943 and activated on 1 June 1943. It was inactivated on 25 August 1943 and re-activated on 15 January 1944 as part of the First Air Force. Made up of the 616th, 617th, 618th, and 619th Bombardment Squadrons, it trained in B-25 *Mitchell* medium bombers. Redesignated the 477th Composite Group in June 1945, it was equipped with B-25s and P-47s. It was inactivated on 1 July 1947. For the period 21

above and right
Lieutenant Elwood "Woody" Driver,
99th Fighter Squadron, scored his first
aerial victory over the Anzio Beachhead
in Italy on 5 February 1944. *Lt. Col.
Gene Carter, USAF Retired*

January 1944 through 20 June 1945, the 477th was commanded by Col. Robert R. Selway, Jr., former commander of the 332nd Fighter Group.

Inaugurated by the AAC in response to additional political pressure from leaders of the black community and black press, the 477th began as a paper organization and remained that way until January 1944, when the 332nd was en route to the war in Europe, although planning for its training had begun somewhat earlier. Pilot training class 43-J at TAAF was the first to train cadets in the twin-engine Beechcraft AT-10. Based upon decisions made at the end of basic flight training, about half of the advanced class had 70 hours of multi-engine training before being awarded their wings; combat training would come later at a different location, presumably with and for the 477th. At about the same time, the pilots in the 332nd training at Selfridge learned of the plan to develop a bomber unit. Some of the larger men—of six feet and 200 or more pounds—opted for the bombers believing that fighters, especially the P-39s and P-40s, hadn't been designed with them in mind. Included in the initial group of pilots who were sent to Mather Field, California, for transition training in the B-25 were Lieutenants Bill Ellis, Chappie James, George Knox, Jim Mason, Charles Stanton, Peter Verwayne, and C. I. Williams. After receiving about 100 hours of training in the *Mitchells,* they were returned to Selfridge, where they were joined by the recent Tuskegee pilot graduates for combat training. When the 477th was reactivated in January 1944, it had 60 pilots and no bombardier-navigators or aerial gunners on its rolls; by April of 1945, the Group was still short 26 first pilots, 43 copilots, two bombardier-navigators, and all 288 of its aerial gunners.

In addition to problems faced in developing balanced training programs for black personnel for a bombardment group while providing essential replacements, usually late, for the units already at war, the 477th was handicapped by the attitudes of Colonel Selway, its commander. To describe him as a strict segregationist is an understatement. The change in philosophy and practice espoused by the War Department and assisted greatly in the Air Corps by General Stratemeyer and Colonel Parrish either never reached Colonel Selway's ears or, if it did, made no difference to him or, worse yet, to his superiors in the First Air Force. At Selfridge and later at Godman Field, Kentucky, Walterboro Field, South Carolina and Freeman Field, Indiana, Colonel Selway's practices created such tension and bitterness among black trainees that it is a wonder that it took as long as it did for a major incident to occur.

Among the more blatant examples of the segregationist ways in which the 477th was managed is the fact that black officers—rated and non-rated alike—were treated as though they were less qualified than their instructors, although the reverse was often true. White officers were promoted more rapidly than their black "students" regardless of their qualifications. The same was true of the enlisted force whose technical and administrative training at integrated bases was a nightmare because of the bias they encountered. At Selfridge, housing was allowed to remain vacant because black officers were "trainees" and therefore not entitled to such privileges. WAC and female civilian personnel were escorted to and from work by MPs and warned not to socialize with the black men. Perhaps most galling of all was the designation of recreational facilities for whites and blacks; this flew directly in the face of War Department policy. Colonel Selway, with the support of General Hunter, the First Air Force commander, got around the policy by designating recreational facilities for the sole use of either "instructors" or "trainees." It just happened that all instructors were white and all trainees black. *Jim Crow* by any other name was still *Jim Crow!* The designation of Officers' Clubs for whites and blacks brought the matter to a head shortly after the March 1945 move

Lieutenant "Woody" Driver with a German Junkers transport left behind in Italy. *Lt. Col. Gene Carter, USAF Retired*

from Godman Field (which had proved too small for the 477th's operations) to Freeman Field.

As was his practice, among Colonel Selway's first actions upon arrival at Freeman Field was to designate which facilities were for instructors and which for trainees. Among the buildings so designated were the Officers' Clubs. Officers' Club No. 1, formerly a building for the non-commissioned officers' club, which became known as "Uncle Tom's Cabin," was for trainees. Club No. 2 was for instructors and supervisors. When some of the blacks complained, General Hunter issued orders stating that any insubordination would result in confinement to the guard house, military prison. His racism was revealed in his March 10th statement to the effect that: "I'd be delighted for them to commit enough action that I can court-martial some of them."

The tension, already high for many reasons, burst on the night of 5 April 1945. The 477th Combat Training Squadron arrived at Freeman that day from Godman. Although they had not been briefed on General Hunter's insubordination order, they were aware of the overall tone of segregation that it represented. The black officers had several planning sessions to avoid a violent confrontation and to ensure that the enlisted troops, who were spoiling for a fight, did not become involved. Word that something was going on reached Colonel Selway and he ordered the Base Provost Marshall, Maj. A. M. White, to post men at the front door of Officers' Club No. 2 and to lock the other doors in case any black officers showed up. At about nine in the evening, groups of two or three black officers in full dress uniform began arriving at the white club. When they were politely refused admission, they left quietly—for a while. At quarter to ten, Lieutenants Marsden Thompson and Shirley Clinton arrived at the club and requested admission. When Thompson was refused admission, he stated that he merely wanted to exercise his privilege as an officer in the United States Army Air Forces and to enter the club for a drink. He brushed past the officer at the door and entered the club followed by a group of approximately 18 officers who had accompanied him. Later that evening, another group, led by Lt. Roger "Bill" Terry, followed the same approach and entered the club. A third group entered the club about 30 minutes later. In all cases, entering blacks were placed under arrest by the club officer and confined to their quarters. The same thing happened on the afternoon of 6 April; 24 more black officers were arrested by the club officer, bringing the total to 61 in confinement in quarters. Acting on advice from General Hunter, Officers' Club No. 2 was closed.

On 9 April, the judge advocate general of the AAF ordered the release of all but the three Lieutenants—Thompson, Clinton, and Terry—who had "forced" their way into the club. On that same day, all Freeman Field officers were ordered to sign new Base Regulation 85-2 which prohibited trainee personnel from using the club designated for base personnel and vice versa. This latest of Colonel Selway's tactics had been given First Air Force approval and General Hunter had received expressions of support from the top-most echelons of AAF command. One hundred and one black officers, from the 477th Combat Command Training Squadron and the 619th Bombardment Squadron, refused to sign. Each one who refused was given an individual reading of the 64th Article of War—Mutiny—in the presence of four witnesses, two white and two black, and again ordered to sign. As each refused, he was arrested.

On 10 April, the 101 black officers were flown to Godman Field where they were met by troops from Fort Knox in full battle dress and placed in confinement. Even before they arrived in Kentucky, inquiries began and political pressure started to mount with indications that even the White House was interested in knowing what had really happened. While white officers at Freeman were heard to mutter racially-oriented threats as to what they'd like to do to the blacks, the black officers and enlisted people "kept their cool," maintaining strict military discipline.

Feeling political pressure building, General Marshall, the Army Chief of Staff, ordered the release of all but the three officers who forced their way into the Club. Contrary to implications of AAF responses to numerous Congressional inquiries received, Lieutenants Thompson, Clinton, and Terry were being held for court-martial.

Inquiries and investigations dragged on until July. Reports of the Advisory Committee on Negro Troop Policies of the War Department and of the AAF inspector general were vastly different in their conclusions, with the AAF staunchly defending segregation in the military. Lieutenants Thompson and Clinton were acquitted of their charges and charges against Lieutenant Terry were mitigated to include only the use of force against a provost marshall officer. He was convicted of that charge and fined $50 per paycheck for three months. The best news for the "Freeman Field 101" and for the 477th Bombardment Group was the ordered relief of Colonel Selway and his replacement with Col. Benjamin O. Davis, Jr.,

newly returned from victory in the skies over Europe.

Personnel of the 477th survived charges of a capital crime against 101 of their officers, but the long battle against segregation, personnel shortages in key crew positions, and tensions between black trainees and white instructors had taken an almost insurmountable toll on the morale and efficiency of the unit. Plans to send the 477th to the Pacific, acceptable to General MacArthur, were still being debated by AAF leadership when V-J Day was celebrated.

The treatment of this potentially valuable unit rates only failing marks for the United States Army Air Forces. The 477th Bombardment Group was unable to contribute to the war against fascism abroad, but it scored some victories in the war against *Jim Crow* on the home front!

332nd Fighter Group–July 1944–Victory in Europe

After a very rough start, the month of July 1944 turned out to be a glorious one for the 332nd Fighter Group! In spite of concerns about the 99th's potential impact on the Group, Colonel Davis managed the transition so as to minimize any problems among people and an effective integration of the fourth squadron into the Group began without incident. The 99th and the 100th settled in on either side of the east end of the single runway at Ramitelli and the 301st and 302nd located on either side of the west. Group operations was set up in the large building that had once belonged to an Italian landowner.

The steel runway was quite short and the Group's pilots became proficient at taking off in the shortest distance possible, especially when heading east over the Adriatic. With four squadrons flying, a sixty-eight plane takeoff was not uncommon. As Lieutenant Sawyer pointed out, however, there were losses on takeoff and following aircraft just had to continue their departures through dense smoke and flames.

Transition to the P-51 cost the Group two of its experienced pilots. Captain Mac Ross, a member of the first class at Tuskegee, was killed in a crash on 11 July as was Captain Leon Roberts, the 99th's Operations Officer, who had already accumulated 116 combat sorties without a scratch.

Excessive numbers of missions flown, in comparison at least to their white counterparts, had become a problem for the 99th. Captain Marchbanks, the Group Flight Surgeon, felt compelled to ground a number of pilots until they could build themselves up physically and psychologically. Even General Eaker

Lieutenant Colonel Benjamin O. Davis, Jr. was appointed Commander of the 332nd Fighter Group in October 1943, the same month that the Group was initially alerted for overseas movement. *U.S. Air Force Museum*

replacements for the 332nd and training needs of the 477th exceeded the capacity of training programs at Tuskegee and the AAF wasn't about to open additional training facilities or to train black pilots at integrated flying schools. On the "return from combat" side of the equation, there weren't enough non-combat positions for black pilots to return to unless they were to serve as instructors at Tuskegee. That, too, flew in the face of permitting only white instructors to train black military pilots. The problems were of USAAF creation and the men of the 332nd paid the price! As Gen. Chappie James later put it: "Separate is never equal."

The glory days of July began on the 12th. Planes of the 100th, 301st, and 302nd escorted B-17s of the 5th Bombardment Wing on a raid against railway yards in southern France. The bomber formation was attacked by some 30 FW-190s while the escort was moving into position. Captain Joseph Elsberry of the 301st, fighter group leader, quickly turned into the attacking fighters and ordered the *Mustangs* to drop their tanks. Diving with greater speed and maneuverability, Elsberry picked out an enemy fighter and opened up on him. The German fighter exploded and headed for the ground. Elsberry got a second with a deflection shot and followed a third through a series of split 'S' maneuvers until the German "ran out of sky." Scoring the Group's first triple kill in one day, Captain Elsberry was joined by Lt. Harold Sawyer, also of the 301st, who scored his first victory, downing another FW-190; four enemy aircraft downed and no bombers lost to enemy fighter activity! A great start for a tradition that was to give the "Red Tails" a marvelous reputation among the bomber crews for whom they flew escort.

On the 15th, the Group's four squadrons flew together for the first time on a mission to raid the railway yards of Avignon in occupied southern France. They were returning from a fighter sweep in the area of Vienna, Austria, the following day when they saw an Italian Macchi 205 moving in to attack a lone, straggling B-24. The Italian fighter dove away from the Red Tail's flight leader, but the wingman, 2nd Lt. William Green, turned inside the enemy and, firing almost continuously, followed him down to watch the Macchi crash into a mountainside. Not to be outdone, Capt. Alfonso Davis, flight lead, spotted another Macchi well below his flight level and dove on him with guns blazing until the Italian fighter spun in. Markings on the two enemy aircraft were not recorded, but since the victories came long after Italy's surrender, they are assumed to have been fighting for the

was aware of the problem, noting to General Giles of the Air Staff: "In the recent past, it has been necessary to require the colored pilots to fly more missions prior to retirement from combat than the white pilots in corresponding fighter units in order to retain the 332nd at the same effective fighter strength."

The causes of the problem were twofold, both related to the practices of segregation. The need for

Aviazione Nationale Republicana, Mussolini's unofficial state in the north.

Three Bf-109s fell to pilots of the 302nd—Lieutenants Luther Smith, Robert Smith, and Laurence Wilkins—when they tried to attack elements of the 306th Bombardment Wing over Avignon on 17 July. In his July report, the 302nd Fighter Squadron's historian noted, under "New Fighter Tactics": "Variations in tactics by German fighters were observed by pilots of the 302nd during the month of July, while providing escort for *Liberators* and *Fortresses*. On several occasions, large numbers of Bf-109s and FW-190s were noticed flying in American type formations, on level or 2,000 feet above the bombers. Because of similarity of the formations, the enemy fighters were, on some occasions, thought to be friendly. They would continue in formation until the P-51s or P-38s were out of the way and commence attacks on the bombers in strings of two or three. Another noticeable tactic of the enemy fighters was the practice of making lightning attacks on the bomber formation in strings of two from below and proving adept at performing a split 'S' and heading for the deck. This maneuver proved disastrous for some enemy aircraft."

July 18th proved to be one of the biggest days of the war in terms of "kills" by pilots of the 332nd. Sixty-six *Mustangs* departed Ramitelli to escort bombers of the 5th Bomb Wing; their destination, the Luftwaffe base at Memminger, Austria. In the Udine area, almost three dozen Messerschmitts attacked the bombers, approaching in elements of two and five from three o'clock high and five o'clock low. Again, however, the German pilots seemed inexperienced and somewhat hesitant. Men of the 332nd destroyed nine Bf-109s en route to the target.

Over the target area itself, the Germans sent up another three dozen fighters, Bf-109s, FW-190s, and Me-210s, but they did not attack, flying alongside the bombers but out of range. When four Focke-Wulfs did attack, two were destroyed and the others fled. The tally for the day: 11 enemy fighters downed, three 332nd pilots lost, no bombers lost to enemy air action. The 100th Fighter Squadron was the high scorer as 2nd Lt. Clarence "Lucky" Lester bagged three Bf-109s and 1st Lt. Jack Holsclaw accounted for two more. The 99th's Capt. Edward Toppins downed one of the two FW-190s, his second victory, and his squadron mate, 1st Lt. Charles Bailey, was credited with the other. The 302nd shot down four Bf-109s, one each by Lieutenants Lee Archer, Weldon Groves, Roger Romine, and Hugh Warner. Here the deceptive tactics of the apparently

inexperienced Luftwaffe pilots, reported in the 302nd's history, failed to draw the escorting 332nd fighters away from the bombers. Colonel Davis' very specific instructions never to leave the bombers unprotected backed up by severe penalties, were carefully heeded.

Two days later, aircraft of the 99th, 100th, and 301st escorted B-24s on a raid on Friederichshafen, Germany, and accounted for five more enemy fighters—a fourth for Elsberry and a third for Toppins, as well as initial victories for Captain Armour McDaniel, 301st, and Lieutenants Langdon Johnson and Walter Palmer, both of the 100th.

Unfortunately, three B-24s were lost to flak as the armada proceeded through the very heavily defended Brenner Pass. On a positive note, the "Red Tail Angels" discovered two badly damaged bombers in the Udine area and carefully escorted them home.

German fighters seemed more aggressive on 25 July, when the 332nd escorted bombers raiding the tank factory at Linz, Austria. Sawyer scored his second victory, this time a Bf-109, but the Group lost two *Mustangs* to German pilots. The next day, pilots of the 99th and 302nd killed four more Bf-109s while escorting bombers of the 47th Bomb Wing against Markendorf Air Field in Austria. Toppins got his fourth victory and Romine his second while

Lieutenant Henry "Herky" Perry, 99th Fighter Squadron, and *Apache II*. Postwar, Lieutenant Perry became one of the jet pilot instructors in the newly-integrated Air Force at Williams AFB, Arizona. *Herman "Ace" Lawson, U.S. Air Force Museum*

"Jet Instructors At The Fighter School 1949"

At the leading edge of integration of the USAF, four Tuskegee Airmen entered the Jet Pilot Training Instructor Program For John "Mr. Death" Whitehead, please see page 120.

Vernon Vincent "V.V." Haywood

Vernon Vincent "V.V." Haywood attended Hampton Institute, Virginia, and was accepted into Aviation Cadet Program at Tuskegee Army Air Field. Born in Raleigh,

North Carolina, Haywood arrived at Tuskegee's Chehaw Station in August, 1942, Class 43-D. Sent to Selfridge Field, Michigan, he served with the 302nd Fighter Squadron. As he progressed from Flight Commander and Operations Officer to Commander, he completed 70 combat missions and 356 hours.

Returned to Tuskegee AAF in 1945, Haywood became the Assistant Director, Instrument Training School until assigned to Lockbourne Air Force Base, Ohio. In

1949, as shown in La Grone's painting, Haywood transitioned from piston-engined fighters to train as an instructor in jet-powered aircraft at Williams Air Force Base, Arizona. While there, he served as an Assistant Section Commander and Section Commander in the 3525th Pilot Training Wing. In the United States, Haywood served in Kirtland Air Force Base, NM; Stewart Air Force Base, NY; and at Davis-Monthan Air Force Base, Arizona. He served overseas at Chitose, Johnson and Yokota Air Bases, Japan; at Clark Air Force Base, Philippine Islands and in Saigon, Vietnam at Headquarters MACV. With 29 years of service and having served in three wars, V.V. retired October 1971 as a Colonel, a Command Pilot, having accrued more than 6,000 flying hours, mostly in fighters.

V.V. and his wife, the former Alma Walden, have one son, Vernon Vincent, Jr. Haywood's decorations include: Legion of Merit, Distinguished Flying Cross, Air Medal with 4 Oak Leaf Clusters, Joint Service Commendation Medal, Outstanding Unit Award, Euro/African/Middle Eastern Campaign Medal with 3 Battle Stars, Korean and Vietnam Service Medals, and the AF Longevity Service Ribbon with 5 Oak Leaf Clusters.

V.V. Haywood served ably and well, as a combat pilot in piston-powered aircraft and jets. Honored for his accomplishments and his contributions in three of this country's wars, he was elected to the Arizona Aviation Hall of Fame—a well-deserved tribute.

Henry B. "Herky" Perry

Henry B. "Herky" Perry, Thomasville, GA, was born the youngest of six children of the Reverend Robert N. and Mary A. Jackson Perry. Perry completed flight training in Class 42-H, September 1942, Tuskegee Army Air Field. He was included in the initial cadre of pilots assigned to form the 332nd Fighter Group, Selfridge Air Base, Michigan, and was sent as a replacement pilot to join the 99th Fighter Squadron in Licata, Sicily.

A veteran of twenty-eight voluntary years as a member of the USAF, Perry was credited with 145 combat missions, two and one half aerial victories and 5,500 flying hours. After World War II, he was a fighter pilot, Alaskan Air Command; Commander and Deputy Wing Commander, Thule Air Force Base, Greenland; and Command Staff Operations Officer, 13th Air Force, Clark Air Force Base, Philippines. He also returned to Tuskegee AAF to become Director of Single Engine Advanced Training and an Air ROTC instructor at Tuskegee Institute.

As one of the four depicted by La Grone, Herky Perry reported to Williams Field in Arizona and became a flight instructor and Director of Training and Analysis. In subsequent assignments, he rose to the rank of Colonel and served as Director Combat Operations Center, Central Air Defense Force, Richards-Gebaur Air Force Base, Missouri and Director, Group Operations/Chief, Weapons Division First Air Force, Stewart Air Force Base, Newburgh, New York.

Perry was awarded the Distinguished Flying Cross and Air Medal with six Oak Leaf Clusters. He also received the Air Force Commendation Medal with four Oak Leaf Clusters and the Army Commendation Medal.

Lewis "Lew" Lynch

Lewis "Lew" Lynch attended Ohio State University before entering military service as an Aviation Cadet in 1942. After pilot training at Tuskegee, he graduated in June, 1944, was commissioned an officer and assigned to Walterboro Army Air Field, South Carolina. Lynch completed combat training and joined the 100th FS, 332nd FG, 15th Air Force in Italy. Piloting P-51s during World War II, he flew 42 combat missions.

Reassigned first to Tuskegee and then to Lockbourne Air Force Base, Ohio, as a fighter pilot and flight leader, his separation from the service and return to Ohio State in 1947 was followed by a recall to active duty! He was reassigned to the 100th FS flying P-47s.

Lynch became one of several outstanding pilots of the 332nd who progressed into jet fighters—F-80s, T-33s, F-84s, F-94s and F-100s. He served at Williams Air Force Base, Arizona; Hamilton Air Force Base, CA; Eielson Air Force Base, AK; and Bunker Hill (Grissom) Air Force Base, IN. During his assignment in Alaska, he flew on the Alaskan Air Command's Official Aerial Demonstration Team in an F-86F Sabre jet.

Lynch completed Air Traffic Controllers and Procurement Officers Schools and was assigned to Military Airlift Command (MAC), Scott Air Force Base, Illinois. Out of the military in 1964, he remained as a Contract Specialist, Headquarters MATS, until retirement in the grade of GM-13 in 1985. An Air Medal with three Oak Leaf Clusters, the Presidential Unit Citation and the Air Force Commendation Medal were awarded to him. As a civilian, he received, among others, the Commendation for Meritorious Civilian Service, Pride Achievement Award, Letters of Commendation and Certificates of Achievement.

Married to the former Wallette Coles Bolden, the Lynches have four children and live in University City, Missouri.

Of the La Grone painting of the four selected to be jet instructors, Lew Lynch said, "I was very proud of the fact that Roy chose to include me in his fantastic art work." La Grone was more than proud of the four that he depicted—those who moved from piston to jet power—those who were at the leading edge of integration of the USAF.

John Whitehead see page 120

Captains Harold Sawyer, 301st Fighter Squadron, and Milton Brooks, 302nd, are awarded Distinguished Flying Crosses by then-Colonel Noel Parrish, Commander of Tuskegee Army Air Field. *Harold Sawyer*

Lieutenants Freddie Hutchins and Leonard Jackson were each credited with their first.

On 27 July, the 332nd provided fighter escort for a 47th Bomb Wing raid on the Weiss Armament Works in the Budapest area of Hungary. Despite active enemy opposition, the 332nd knocked down eight more fighters—four FW-190s and four Bf-109s. First Lieutenant Ed Gleed of the 301st accounted for two of the FW-190s and his Squadron mate, 2nd Lt. Alfred Gorham, got the other two. 1st Lt. Leonard Jackson of the 99th shot down his second Bf-109 in two days and Capt. Claude Govan of the 301st shot down his first. Initial kills were also recorded by Lt. Richard Hall of the 100th and Lt. Felix Kirkpatrick of the 302nd. The final victory of the month was recorded by 2nd Lt. Carl E. Johnson of the 100th Fighter Squadron when he shot down an Italian Reggiane 2001, another of those mysterious fighters presumed to be from northern Italy.

All things considered, July was extraordinary for the 332nd Fighter Group. The weather was good,

they were settling into their new home near Ancona on the east coast of "the Boot," and they were performing a key mission very successfully. Although losses were sustained—in transition to the P-51 and to the enemy—the morale of the entire organization had been bouyed by the almost incredible total of 39 victories! And, they had protected their "heavies" against enemy fighters! The "Red Tails" were coming into their own.

Good summer weather permitted 332nd Fighter Group participation in a range of missions once thought impossible for black pilots. They flew against hard targets in southern France, against Austria and Germany itself, and against the infamous oil refinery at Ploesti, Rumania, where, unlike so many other units, they suffered no losses. They flew against targets throughout the Balkans and against enemy installations in Yugoslavia, where they took a number of temporary losses; in a number of cases pilots forced to bail out or to crash-land were rescued by Yugoslav partisans and helped to return to their units. Those returns sometimes took as long as a month of hazardous travel through mountainous terrain and enemy-held territory.

Good weather also helped a variety of recreational activities. Squadron clubs were built and enjoyed; the 99th's *Panther Room* reflected added "procurement" experience of the Squadron, but all provided the hard-working enlisted force with places to "wind down" away from the flight line and the officers. Sports competitions were seasonal and active with the 99th's baseball team coming in first in the base league, closely followed by the 301st. The Adriatic's warm waters at nearby beaches were missing only hot dog stands and popcorn vendors to be truly like home. Visits to nearby USO shows and to R&R (rest and recreation) camps at Rome, Naples, and Montesoro provided welcome breaks and a visit by S/Sgt. Joe Louis, the heavyweight champion, was a definite highlight. Morale throughout the Group was high as confidence in themselves and in the ultimate victory grew.

Aerial victories in August did not rival the extraordinary July numbers, but five more pilots scored kills: 2nd Lt. George Rhodes of the 100th shot down an FW-190 and Flight Officer William Hill, 302nd, bagged a single Bf-109. On 24 August, on an escort mission with the 5th Bomb Wing to a Czech airfield, three enemy fighters fell to the guns of the "Red Tails": Lt. John Briggs of the 100th knocked down a Bf-109 and Lieutenants Charles McGee and William Thomas of the 302nd each got an FW-190.

Enemy opposition in the air seemed greatly diminished by the end of August and the 332nd had two field days attacking Jerry aircraft on the ground. The first came on the 27th on a strafing mission to the Grosswardein Airdrome.

The Group Historian wrote: "Our pilots strafed the German planes by squadrons. Not a single enemy rose to meet them in an effort to stop the destruction, nor was any anti-aircraft fire directed toward the vaunted *Mustangs* as they swept over the field like a cloud of grasshoppers over ripened western grain."

Captain Alfonso Davis, the Deputy Group Commander who led the mission, said: "They were parked there like ducks. All we had to do was shoot them," and Lt. Bernie Jefferson suggested, "They must have lost all discipline."

A similarly successful attack took place on August 30th when the Group returned from an escort mission to Blechhammer, Germany. Having brought the bombers back to safety, the flight lead, Capt. Melvin Jackson, spotted an enemy aircraft

just lifting off. Diving down to investigate, he located two adjacent airfields, at Kostelleo and Prostejeu, Czechoslovakia, and the Group went to work on the target-rich environment. Twenty-two enemy aircraft were destroyed and 18 more were damaged bringing the month's total to 105, according to the Group report.

Unopposed attacks on enemy aircraft on the ground continued into September. On the 8th,

Captain Harold Sawyer, 301st Fighter Squadron, scored his first aerial victory on 12 July 1944 on a mission to southern France and his second just 13 days later on an escort mission to Austria. *Harold Sawyer*

Capt. William Mattison led an attack against Ilandza Airdrome in Yugoslavia. Eighteen enemy aircraft—Focke-Wulfs, Junkers, Heinkels, and Dorniers—were left burning. En route home to Ramitelli, the flight spotted more aircraft lined up on the ramp at the Alibunar Airdrome, also in Yugoslavia, and 18 more were destroyed.

McGee, one of the flight leaders, reported: "With our planes functioning perfectly and statue-like targets so choicely placed, we couldn't miss."

August and September 1944 brought good news to the 332nd in many forms. Five pilots reported as missing in action returned! Second Lieutenant Robert McNeil returned after spending some time with the Free French Maquis in southern France and Lt. Charles Jackson returned after more than a month in Yugoslavia. Three more came back in September after having been forced down in enemy territory, their stories carefully listened to by their comrades-in-arms! Promotions for both officers and enlisted members recognized effort and experience and awards and decorations gave recognition of heroism and extraordinary achievement. A September 10th ceremony was a special highlight for the entire Group! Colonel Davis, Captain Elsberry,

Harold Sawyer, on leave in December 1944, with then-Ohio Governor Frank Lausche. *Harold Sawyer*

Lieutenant Leon Roberts, 99th Fighter Squadron. Time out in Italy. *U.S. Air Force Museum*

and 1st Lieutenants Holsclaw and Lester were presented Distinguished Flying Crosses by Colonel Davis' father, Brig. Gen. Benjamin O. Davis, Sr. Lt. Gen. Ira Eaker, Commander of the Mediterranean Allied Air Forces; Maj. Gen. Nathan Twining, Commander, 15th Air Force; and Brig. Gen. Dean Strother, Commander of the 15th Fighter Command, attended to add their congratulations and to participate in the Review.

Among August's new arrivals was Maj. Spanky Roberts, returning to command a very happy 99th Fighter Squadron for the third time! In spite of some truly bad weather and delayed mail, morale remained high, bolstered by continuing victories, by good news on all war fronts, and by the return of those thought to have been lost.

After "highs" experienced in July, August, and September, reviews of October 1944 were "mixed" at best. The highest losses of the war were suffered, with sixteen pilots killed or missing in action; five each from the 99th and 100th and three each from the 301st and 302nd. Three of those missing in action returned to safety in a relatively short time and hope for the rest was very much alive.

On the plus side, the Group had another big aerial victory day on 12 October. Attacking river traffic on the Danube for a second consecutive day, each of the four squadrons engaged different targets; the 302nd was the only one to encounter hostiles in the air. Captain Wendell Pruitt and Lt. Lee Archer had a field day, three Bf-109s downed by Archer and one He-111 and another Bf-109 by Pruitt. In addition, Capt. Milton Brooks and Lieutenants Green, Romine, and Luther Smith each shot down one enemy aircraft. The "Red Tails" destroyed 26 enemy aircraft on the ground as well as a large number of trucks and river barges. Coupled with the bombers and transports destroyed in raids on Greece earlier in the month and those destroyed on the Blechhammer raid on the 13th of October, 47 enemy war planes were removed from service against the Allies. The missions for the rest of the month were focused on oil installations and against targets of opportunity. None were especially noteworthy although the Group maintained its reputation of never losing a bomber to enemy air activity.

Five experienced pilots completed their tours of duty overseas and were returned home for rest and reassignment. Among those departing for the United States was Pruitt of the 302nd, to be long remembered for his victory over the German destroyer and his "buzzing" of the operations building at Ramitelli.

The senior half of "the Gruesome Twosome" with Lee Archer, Pruitt returned to Tuskegee with two Messerschmitts and one He-111 officially credited.

"Winterization" had become a major effort by the time early November rolled around. Officers and men of the 332nd were reluctant to face another muddy, cold, windy, rainy winter without having done all that they could to protect themselves from the elements. Captain Ray B. Ware, the Group Historian, wrote: "The sound of hammer on nail, and sometimes on finger, was heard and gradually an area of tents became transformed into an area of tents and houses, semi-tents and semi-houses, and in some cases among the tents and houses, a mansion appeared. …To see an ordinary tent supported by tent pegs and side flaps was an item of curiosity."

Another activity occupying a good deal of attention was the carefully staged production of the Group play, "G.I. Joe in a Vino Joint," reported to have been a sophisticated comedy. The touch football league was popular and apparently violent with "the crash clashing of bodies…heard from one end

Lieutenant Leon Roberts, 99th Fighter Squadron, repacking a parachute—"just in case". Doing it yourself helped your confidence. *U.S. Air Force Museum*

James
E. P.
Randall

Colonel James E. P. Randall, USAF Retired, is a native of Roanoke, Virginia. He attended Hampton University prior to entering pilot training in the U.S. Air Force. Having completed basic training at Randolph Air Force Base, Texas and advanced at Las Vegas, Nevada, Randall became a flight instructor at Perrin and Craig Air Force Bases in Texas and Alabama.

Randall flew 75 missions over North Korea piloting F-51s during the Korean War. Between 1953 and 1957 he made eight jet fighter crossings of the North Atlantic Ocean.

In F-100s, he was a gunnery instructor at Nellis Air Force Base, Nevada and, as a F-100 pilot, commanded the Alert Facility in Spangdahlem, Germany.

In F-105s, then-Major Randall was assigned on temporary duty to South East Asia. On his forty-fourth mission over North Vietnam, he was shot down. Rescued, he was returned to the United States for medical treatment.

Randall was assigned to Nellis Air Force Base, Nevada as Operations Officer for the testing of the swept-wing F-111 fighter. It is as a test pilot that La Grone depicted Randall. Randall served for thirty-one years, accrued more than 7,200 flying hours in more than 24 conventional and jet aircraft. He made a total of sixteen crossings of the Atlantic and Pacific Oceans in jet fighters.

At Tyndall Air Force Base, Florida, Randall served as Base Commander. He and his wife Essie reside in Colorado Springs, Colorado where he retired as Director of Operations Plans and Chief of Safety, Headquarters Air Defense Command. His many decorations include the Legion of Merit, Distinguished Flying Cross, Bronze Star, Meritorious Service Medal, Air Force Commendation and Air Medal and the Purple Heart.

of the field to the other" and four boxers from the Group reached the finals of the 15th Air Force Boxing Tournament at Foggia. Basketball practice began and the anticipated arrival of text books speeded up plans for educational opportunity.

The focus on ground activity and preparation for winter was brought on by overcast skies that resulted in a great many "stand-down" days and frustration for the fliers. Captain Luke Weathers of the 302nd recorded the only aerial victories in either November or December. In November, Captains Weathers, Melvin Jackson, and Louis Purnell were escorting a lone bomber to safety after a mission to Munich, Germany, when they were attacked by a flight of eight Messerschmitts. Although badly outnumbered, Weathers turned directly into the enemy fighters' path and, firing short but lethal bursts, knocked down one Bf-109 right away. Then, under attack from his "six o'clock" position, he chopped power and lowered his gear and flaps to slow down abruptly. The trailing German overshot and Weathers was on his tail firing short machine-gun bursts until the enemy fighter exploded and fell to earth. The bomber returned safely to its base in Italy and "Red Tails" tradition was upheld.

Other bomber escort missions to Germany, Austria, and Czechoslovakia were flown, although in reduced numbers because of the weather, and December 9th provided the first glimpses of the newest German weapons system, the Messerschmitt Me-262 jet fighter. In separate sightings, one southeast of Regensberg and the other east of Munich, two Luftwaffe jets seemed out to demonstrate the high speed capabilities and maneuverability for which they were to become noted, without engaging in actual combat—at least not yet.

Although it had become apparent that the Germans, with their greatly shortened supply and communications lines, were going to defend their homeland vigorously and that the end of the war was not going to be before Christmas 1944, morale in the 332nd Fighter Group had never been better. There were losses in each of the four Squadrons during both months; however, they were, to some extent, offset by the returns of a number of fliers who had been listed as missing in action.

A number of veteran pilots left to return home in time for the holidays even as their replacements arrived. Promotions and awards for officers and enlisted personnel were almost as big a morale boost as the steady arrival of holiday mail. Thanksgiving was celebrated with the traditional feast of turkey

Who's up next? Crew Chiefs Tommy Downs and Alexander Crawford survey the status board. *U.S. Air Force Museum*

their climax, Colonel Davis returned from his trip to the United States; his arrival and his Christmas Day "interviews" were boosts to the already soaring morale. His year-end 1944 message to the officers and men of the 332nd was clearly and quietly indicative of the job they had done: "A definite milestone has been passed in our history. When we embarked for foreign service, we were completely untried. Since that day the Group has taken on the many varied assignments given it and carried them out with distinction and success. I cannot fail to mention the all important fact that your achievements have been recognized. Unofficially you are known by an untold number of bomber crews as the Red Tails who can be depended upon and whose appearance means certain protection from enemy fighters. Bomber crews have told others about your accomplishments, and your good reputation has preceded you into many parts where you may think you are unknown. The Commanding General of our Fighter Command has stated that we are doing a good job, and that he will so inform the Air Force Commander. Thus, the official report of our operations is a creditable one."

The Group Historian noted in January: "While generally a cold dark grey murkiness prevailed during most of the month, the "Red Tails" did find opportunity to register another eleven missions against the enemy. Diversified missions and targets included photo reconnaissance and bomber escort missions to communications and oil targets at Vienna, Munich, Praha, Stuttgart, Regensburg and Linz. The Luftwaffe, with seemingly no reluctance, continued to relinquish its rights to the air, thereby utterly failing to provide even one item as a topic for the special accounts column in January '45's Chronology."

Without many missions to fly, the Group again concentrated on its ground training for pilots and technicians. A pilot/crew chief roundtable was organized and the arrival of a P-51 Mobile Training Unit, after six months flying the airplane, still provided new insights into mechanical operations of the *Mustang*. Two ceremonies were held to recognize bravery by awarding Distinguished Flying Crosses, Bronze Stars, and Air Medals. Thirty-four replacement pilots arrived and began to assume their places in the fighter squadrons. In spite of the cold, rain, wind, and mud, morale remained high and the Russian army's steady progress on the eastern front was noted for its role in hastening the war's finale.

Surprisingly, February 1945 offered greatly improved weather across the continent and the

and trimmings in each mess hall and with chapel services and movies. In the traditional "Turkey Day Classic" touch football game, the 99th defeated the Group Headquarters team 12-0. Each of the squadrons' musical groups practiced and performed and the Group newspaper made its debut.

Among the very welcome commendations received at year end was that of Gen. Nathan Twining, 15th Air Force Commander, who praised the untiring efforts of the aircraft maintenance crews in keeping the aircraft operational. As Christmas celebration activities were reaching

North American P-51 Mustangs of 332nd Fighter Group, 15th Air Force, Italy. The famed "Red Tails" of the 332nd never lost an escorted bomber to enemy fighter attack. *Roy E. La Grone*

Lloyd W. "Fig" Newton

Major General Lloyd W. "Fig" Newton—a fortunate recipient of the legacy of the Tuskegee Airmen—is currently Director of Operations, J3, U.S. Special Operations Command, MacDill Air Force Base, Florida. He assumed that position in July, 1993.

In 1974, Fig Newton became the first black to perform as a pilot with the U.S. Air Force Aerial Demonstration Team, The Thunderbirds. He took advantage of the doors that had been opened by those that flew in World War II and widened them. He held several positions with the elite performance group: narrator, advance coordinator, slot pilot (1976-77) and right wing (1978).

A native of Ridgeland, South Carolina, General Newton graduated from Jasper High School in 1961. The holder of a bachelor's degree in aviation education from Tennessee State and a master's degree in public administration from George Washington University, he also completed three military post-graduate courses; Armed Forces Staff College, Industrial College of the Armed Forces and Harvard University's National Security Senior Executives Course.

A distinguished graduate of the Air Force Reserve Officer Training Corps program, General Newton completed pilot training at Williams Air Force Base, Arizona, in 1967. He was assigned to George Air Force Base, California, to undergo training in the F-4 Phantom preparatory to service in Vietnam. As a pilot during the war in Vietnam, General Newton flew 269 combat missions, including 79 over North Vietnam, from Da Nang Air Base.

His tour in Vietnam over, General Newton was assigned to the 405th Tactical Fighter Wing, Clark Air Base, Philippines. Returning to the United States in 1974, he became an instructor in F-4s at Luke Air Force Base, Arizona, and then one of the legendary Thunderbirds. His was an auspicious first. He had, as mentors, the equally legendary Tuskegee Airmen.

At the completion of his tour with The Thunderbirds, General Newton transitioned into F-16 fighters at MacDill Air Force Base, Florida. He was then assigned to the 8th Tactical Fighter Wing, Kunsan Air Base, South Korea, as assistant deputy commander for operations, and then to the 388th Tactical Fighter Wing, Hill Air Force Base, Utah.

A command pilot with more than 4,000 flying hours, General Newton has commanded the 71st Air Base Group and 71st Flying Training Wing, both at Vance Air Force Base, Oklahoma. He commanded the 12th Flying Training Wing, Randolph Air Force Base, Texas, and the 833rd Air Division and the 49th Fighter Wing, both at Holloman Air Force Base, New Mexico. He has been decorated with the Legion of Merit with oak leaf cluster, Distinguished Flying Cross with oak leaf cluster, Meritorious Service Medal with oak leaf cluster, Air Medal with 16 oak leaf clusters, Air Force Commendation Medal, Air Force Outstanding Unit Award, Philippines Presidential Unit Citation, Vietnam Service Medal and Republic of Vietnam Campaign Medal.

332nd launched a record total of 141 missions, principally against German rail traffic. Strategic raids against key oil production facilities had taken their toll and the enemy was forced to rely on a railroad network to move troops and vital supplies. One particularly noteworthy mission was flown by 45 *Mustangs* from Ramitelli against targets in the area bounded by Munich, Ingolstadt, Linz, and Salzburg. Led by Group Operations Officer, Capt. Ed Gleed, the "Red Tails" destroyed locomotives and a power station and damaged numerous freight cars. In a strafing pass at an airdrome in the area, one pilot destroyed a Bf-109 fighter, two Heinkel bombers on the ground, and damaged a third.

In addition to the vastly increased and highly successful aerial activity, the Group history includes notations on: A wave of sentiment over Valentine's Day and little recognition of two Presidential birthdays; results of the "Miss 332nd Pin-Up" contest and the prizes sent to Miss Betty Gassoway at the University of Chicago; dispersal of the men of the 302nd Fighter Squadron to the three other units in anticipation of the 7 March 302nd's de-activation.

March was characterized by heavy personnel turnover in the three squadrons continuing war against Hitler, pilots returning home, new pilots arriving, and the inclusion of experienced pilots and ground crew from the 302nd. Sadness over losses to the enemy and the departure of comrades was eased by large numbers of well-earned promotions and awards to officer and enlisted personnel. It was also characterized by a return to the strategic bomber escort role and to air-to-air combat.

The big difference in the aerial combat scene was the entrance of the Me-262 jet as a real combatant. Fifty-nine P-51s departed Ramitelli on 24 March to escort B-17s of the 5th Bomb Wing on one of the 15th Air Force's longest missions. The target: the Daimler-Benz tank works in Berlin. In addition to providing cover for the bombers, the 332nd was expected to draw the attention of the few remaining Luftwaffe pilots away from airborne landings in the area north of the Ruhr.

The Group expected to be relieved upon reaching the enemy capitol, but the relief fighters were late to the rendezvous point and the "Red Tails," faithful to their reputation, stayed with the "heavies." Just after noon, the formation was attacked by the German jets and it took the American fliers a moment to figure out the high speed, shallow turn tactics being employed.

Lieutenant Herbert E. "Gene" Carter, Chief of Maintenance and fighter pilot, 99th FS nicknamed "Pepi Le Moko," with his P-40 named for his wife, Mildred, who was nicknamed "Mike."
Lt. Col. Gene Carter, USAF Retired

Dr. Roscoe Brown recalled, "They were coming in, climbing to hit the B-17s from nine o'clock. I made a reverse peel to the right, dove underneath the bombers, broke out to the left. They never saw me coming up at them."

Several bursts from the *Mustang* found their mark and the German pilot bailed out to watch his aircraft hit the ground. Thus, Lieutenant Brown became the first of the 332nd to shoot down a German jet. He was joined quickly by two other pilots of the 100th—Flight Officer Charles Brantley and Lt. Earl Lane—in this distinction. Of the eight Me-262s downed, the 332nd was credited with three.

The second jet fell to Brantley. He saw his shells hit the Messerschmitt fuselage and he saw it nose over and head down. But he didn't see it crash; the other pilots did and his kill was confirmed. Lane bagged the third after a furious diving chase that started around 20,000 feet and ended when the heavily smoking jet crashed into the ground. For

each of the three, it was the first victory. For the Group, their heroic actions against the enemy and their faithfulness to the bombers were recognized in the award of a Distinguished Unit Citation.

March 31st and April 1st, 1945 could be described as "putting the icing on the cake" of the Group's aerial combat experience. On the first of those two days, the pilots of the 99th and 100th, in a record performance for the 332nd, shot down thirteen enemy aircraft in a mass dogfight over Linz, Austria. Seven Bf-109s and 6 FW-190s fell to guns of the "Red Tails." Three more "probables" were noted into the records and there were no American losses! First Lieutenant Robert Williams accounted for two of the Focke-Wulfs and Brown and Lane each added a Bf-109 to the Me-262 destroyed on the 24th.

Not to be outdone by their compatriots, the men of the 301st had their own "turkey shoot" the following day. Seven pilots shot down twelve German fighters in the area around Wels, Austria. First Lieutenant Harry Stewart joined an elite group of "Red Tails" when he bagged three FW-190s on a single mission. First Lieutenant Charles White

knocked down two Bf-109s as did 2nd Lt. John Edwards, and 2nd Lt. Carl Carey added two more Focke-Wulfs to the total. Lieutenants James Fischer, Walter Manning, and Harold Morris each shot down one FW-190. Twenty-five in two days! Extraordinary!

As the Group Historian reminds us, however, not all the damage was done against aerial targets: "In its record month, the "Red Tail" Group varied between bomber escort, photo reconnaissance escort and strafing missions. In four strafing attacks against the enemy's lines of communications…, the "Red Tails" accounted for the destruction or damage of 69 locomotives, 269 box cars, 37 flat cars, 27 rail passenger cars, 22 oil cars, 13 motor transports, 10 warehouses, 5 railroad stations, one railroad power station and 17 aircraft on an enemy airdrome, as well as many minor installations."

Truly a record period, but not without its sadness. Captain Armour McDaniel, C.O. of the 301st, was reported as missing in action on the Berlin raid and twelve other Group pilots were also MIA. One pilot was killed on takeoff and another died of burns received when a wing tank fell from

John L. Whitehead Jr.

First Black Experimental
Test Pilot in the U.S. Air Force

John L. Whitehead, Jr., attended West Virginia State College prior to entering the U.S. Army Air Corps. He received his pilot's wings and was commissioned a 2nd Lieutenant in September of 1944. He joined the 301st Fighter Squadron, 332nd Fighter Group in March 1945 and flew 19 missions, returning to the United States in the Fall. First assigned to Tuskegee Army Air Field, he was transferred to Lockbourne Army Air Base and separated from the service in January 1947 in order to return to West Virginia State. He graduated in 1948 with a B.S. in Industrial Engineering.

Recalled to active duty, August 1948, Whitehead was assigned to the 100th FS, 332nd FG at Lockbourne. With deactivation of the Group due to integration, he was assigned to Williams Air Force Base, Arizona, as the USAF's first black jet pilot instructor. Then assigned to Boeing Aircraft Corporation, Whitehead served as the Training Command Representative on the B-47 aircraft.

At the outbreak of war, he flew 104 missions in Korea. He served as Chief Production Test Pilot for the Air Force at Northrup Aircraft Corporation and, in 1956, was reassigned to Hill Air Force Base, Utah, as a test pilot.

Whitehead was selected to attend the USAF Experimental Test Pilots School, Edwards Air Force Base, California, from which he graduated in January 1958—the first black experimental test pilot in the Air Force. He returned to Hill Air Force Base, Utah, as Chief of the Quality Analysis Division. He was stationed in Chateauroux, France, and McClellan Air Force Base, California, prior to the war in Vietnam. For the 56th Special Operating Wing in Southeast Asia, he flew combat missions in the C-47 aircraft.

Returned to Edwards Air Force Base, California, Whitehead was named Chief of Standardization and Evaluation Division, then Maintenance Squadron Commander. Committed to excellence, John L. Whitehead, Jr., was popularly known as "Mr. Death" because of his extremely slender build. He retired as a Lieutenant Colonel and Deputy Group Commander of the Maintenance and Supply Group at Edwards Air Force Base. During his thirty-year career, this Tuskegee Hero was awarded the Distinguished Flying Cross with OLCs; Air Medal with seven OLCs and the Air Force and Army Commendation Medals. He logged an incredible 9,500 flying hours with over 5,000 hours in jet aircraft and he flew 40 different types of aircraft including fighters, bombers, and transports.

an overhead aircraft and exploded on his tent. In spite of the losses, however, Group morale continued on the upswing. "The rise of the 'Red Tail' morale is inversely proportional to the fall of the Reich," reported the Historian.

Morale was also boosted by the visit of Truman Gibson, Judge Hastie's successor as civilian aid to the Secretary of War, by the presentation of numerous awards and the announcement of a large number of promotions and by victory in the 15th Air Force basketball tournament.

Beginning as it did, with twelve aerial victories on April Fool's Day, the following month, the last in the War in Europe, brought more glory to the pilots of the 332nd. Lieutenants Jimmy Lanham and William Price each downed a Bf-109 and, while escorting a lone P-38 photo reconnaissance aircraft to Prague, Czechoslovakia, the "Red Tails" scored the final four aerial victories in the Mediterranean Theater. Four Bf-109s were shot

The 477th Bomb Group on Parade. *U.S. Air Force Museum*

Mechanics of the 477th Bombardment Group perform a major inspection on a North American B-25J Mitchell medium bomber. *Dave Menard, U.S. Air Force Museum*

Guion S. Bluford, Jr.

Guion S. "Guy" Bluford, Jr., was the only Air Force member and the first U.S. black to travel into space in 1983, an earlier and successful mission of the ill-fated, 100-ton space shuttle Challenger , later lost in its unforgettable disaster. Bluford thundered skyward in a mission for which he was named a mission specialist by NASA in 1978. "Some people have asked if I climbed aboard the shuttle fearing for my life," Bluford said. "I didn't. Sure, my adrenaline was pumping, but like the others, I was looking forward to this flight. I'm used to the flying environment; I felt very comfortable and very well prepared."

Bluford was tasked with creating a computer program to calculate the flow field around delta wings; wings similar in shape to those of the Shuttle. While a staff development engineer for advanced concepts for the Aeromechanics Division and branch chief of the Aerodynamics and Airframe Branch in the Flight Dynamics Laboratory at Wright-Patterson Air Force Base, Dayton, Ohio, he had written several scientific papers on fluid dynamics.

Experienced as a combat pilot, Bluford earned a Bachelor's Degree in Aerospace Engineering from Pennsylvania State University where he also participated in Air Force ROTC. He completed F-4C combat crew training and, in 1966-67, was assigned to Cam Ranh Bay, Vietnam. He flew 144 combat missions, 65 of them over North Vietnam.

Assigned to Sheppard Air Force Base, Texas, Bluford served as a T-38A instructor pilot and then transferred to Wright-Patterson Air Force Base where he earned his Master's and a Ph.D. in Aerospace Engineering. He studied laser physics through the Air Force Institute of Technology (AFIT) program.

Bluford said to youth: "Hard work, dedication, and perseverance are the keys to success. Work to get an education, and be as prepared as you can possibly be for any opportunities that may come your way. You can achieve any worthwhile goal you set your mind on, if you work hard and persist long enough." From one who knows, he added, "The sky's the limit."

Charles F. Bolden, Jr.

Charles F. Bolden, Jr., USMC, of Columbia, South Carolina, received a Bachelor of Science Degree from the U.S. Naval Academy in 1968, majoring in Electrical Science. A decade later, his Master's Degree in Systems Management was obtained at the University of Southern California.

Trained in the A-6A Intruder at Cherry Point, North Carolina, he flew more than 100 sorties into North and South Vietnam, Laos and Cambodia while assigned at Nam Phong, Thailand. In June 1979, he graduated from USN Test Pilot School and was assigned to the Naval Air Test Center's System Engineering and Strike Aircraft Test Directorates. Bolden served as an ordnance test pilot and flew numerous test projects in A-6E, EA-6B, and A-7C aircraft. Having logged more than 2,600 flying hours, 2,300 in jet aircraft, he was selected as an astronaut candidate in May 1980.

Frederick D. Gregory

Frederick D. Gregory, of Washington, DC, received a Bachelor's degree from the USAF Academy in 1964 and a Master's Degree in Information Systems from George Washington University in 1977.

Gregory flew helicopters for three years, including a tour in Vietnam as a rescue crew commander. Having completed U.S. Navy Test Pilot School, he was a research/engineering test pilot for the Air Force at Wright-Patterson Air Force Base, Ohio and for NASA at the Langley Research Center, Virginia.

Gregory, USAF, the first black to command a Shuttle Mission, has flown more than 40 different types of single- and multi-engined fixed and rotary wing military and civilian aircraft including gliders. Having logged over 4,100 hours of flight time and holding a commercial and instrument certificate for single- and multi-engine and rotary aircraft, Gregory was selected as an astronaut candidate in January, 1978.

Ronald E. McNair

The late Col. Ronald E. McNair was born in 1950 in Lake City, South Carolina. He received a Bachelor's Degree in Physics from North Carolina A&T State University in 1971 and a Ph. D. in Physics from Massachusetts Institute of Technology (MIT) in 1976. While at MIT, Dr. McNair performed some of the earlier developments of chemical and high pressure lasers. His later experiments and theoretical analysis on the interaction of intense CO_2 laser radiation with molecular gasses provided new understandings and applications for highly excited polyatomic molecules.

A staff physicist with Hughes Research Laboratories in California, he conducted research on electro-optic laser modulation for satellite-to-satellite space communications, the construction of ultra fast infrared detectors, ultraviolet atmospheric remote sensing, and scientific foundations of the martial arts. He was selected as an astronaut candidate in 1979.

Shown in purple, Roy La Grone's choice in that ancient Greeks used the color to signify death, Ron McNair seems larger than life. A victim of the dreadful explosion that destroyed the Challenger and all of the astronauts aboard, January, 1986, Ron McNair is sorely missed.

Dr. Mae Jemison and others have subsequently joined the ranks of U.S. astronauts. In all, they represent the country's very best.

Captain Alfonso Davis, 99th Fighter Squadron, was later killed in a plane crash in Italy, 1944. *Herman "Ace" Lawson, U.S. Air Force Museum*

down by *Mustangs* flown by 2nd Lt. Thomas Jefferson, who bagged two, 1st Lt. Jimmy Lanham, his second victory in eleven days, and 2nd Lt. Richard Simons.

Although the war was clearly winding down, with daily notices of Allied victories on both fronts, and morale should have been approaching a peak, news of the death of the Commander-in-Chief had a particularly harsh impact on the men of the 332nd. President Roosevelt was widely regarded as a true friend of black American aviators.

One last story from the Group Historian: "F/O James H. Fisher of Boston, Mass. returned to base approximately two hours after the mission had landed (on April 1st). He was sans airplane, sans parachute, but he brought with him a tale, which, in its entirety, was more action packed and thrilling than one of Street and Smith's dime novels. He shot one FW-190 down and damaged two others. He was hit twice by fire from enemy aircraft and twice by flak. His controls were shot away and he was forced to fly by his (trim) tabs. His gas drained out through flak holes. He was forced to bail out over Yugoslavia, where he was surrounded by partisans, then taken to Zara. He then boarded a C-47 and landed at his home base shortly after the other members of his flight. All in a days work."

War against fascism in Europe ended on V-E Day, 8 May 1945. Awards ceremonies and parades recognized the valiant efforts of pilots and ground

crew and paid solemn tribute to those who had given their lives in the cause of freedom. Distinguished Flying Crosses and Air Medals rewarded the courage and skill of fliers and Bronze Stars commended the continuous devotion of ground personnel to "keep 'em flying." A team effort that had made significant contributions to the hard-won victory over Nazi Germany was warmly praised by the Group's Wing Commander, Colonel Taylor, and by General Twining, C.O. of the 15th Air Force.

During the first week in May, the Group completed its last move of the war, to Cattolica Airport, where spotless white buildings welcomed the members to their "last campsite." Training was initiated to prepare for a move into the ongoing war against Japan, but the pace was far more relaxed and there was time for leave visits to Venice, Genoa, and Milan. Active recreation programs were established and, coupled with expectations of returning home, they maintained morale at a high level. Naptha University continued to provide educational opportunities for the men and some, including Roy La Grone, enrolled at the University of Florence.

In early June, Colonel Davis was appointed Commander of the 477th Composite Group at Godman Field, Kentucky and he departed Cattolica, almost immediately, by B-17 with 40 members of the Group. His final letter to the men of the 332nd, recorded in Group history, read: "In parting I would like to say that it has been a signal honor to be part of such a fine organization as the 332nd Fighter Group. It is with regret that I leave you with whom I have been so closely associated during the past many months. We have been through much together, and the many common experiences we share make it difficult for me to

tear away so abruptly. However, the war is but half won, and we cannot let up until Japan is defeated. In the words of Brigadier General D.C. Strother who recently commanded our Fighter Command, the 332nd has been a credit to itself and the Army Air Forces."

The wartime part of the Tuskegee Experience ended on 16 October 1945, at Camp Kilmer, New Jersey. The three remaining squadrons were deacti-

A crew chief with a "liberated" Germany motorcycle in Italy. *U.S. Air Force Museum*

Lieutenant Lee "Buddy" Archer and his P-51, Mustang, "Ina the Macon Belle." Archer, and Lt. Wendell Pruitt, 302nd Fighter Squadron, were known as the "Gruesome Twosome"—a deadly duo. *Roy E. La Grone*

Colonel Benjamin O. Davis, Jr., Commander, 332nd Fighter Group, briefs his troops before a long-range bomber escort mission. *U.S. Air Force Museum*

vated and the men were dispersed. Some went home, some went back to Tuskegee, and some went to Kentucky to join Colonel Davis and the all-black 477th. After all that they had done, they were winding up very close to *where* they had begun, segregated from the mainstream of American economic, political, and social life.

They also wound up as *who* they were, courageous and capable black Americans more determined than ever to change their status in the society and to keep contributing as they had so valiantly done throughout their time in combat. The Tuskegee Experiment and the Tuskegee Experience tempered their already fine steel to a strength that has sustained them in all their postwar persuits.

302nd Fighter Squadron Commander, Capt. Edward C. Gleed, briefs 332nd Fighter Group pilots on the mission to be flown. *Official USAF Photo, U.S. Air Force Museum*

Lieutenant Willie Fuller reads a letter, relaxing on the wing of his aircraft. Lieutenant Fuller of the 99th Fighter Squadron was among the first of the Tuskegee Airment into combat. He flew 70 combat missions in North Africa, Sicily, and Italy. *Herman "Ace" Lawson, U.S. Air Force Museum*

CHAPTER FOUR

Epilogue—Toward Tomorrow

"We are returning from war!" wrote W.E.B. DuBois, the intellectual who initiated the Niagara Movement that developed into the National Association for the Advancement of Colored People. In *The Crisis*, the journal he edited for the NAACP, he continued, "This is the country to which we Soldiers of Democracy return. This is the fatherland for which we fought! But it is our fatherland. It was right for us to fight. The faults of our country are our faults. Under similar circumstances, we would fight again. But, by the God of Heaven, we are cowards and jackasses if now that the war is over, we do not marshal every ounce of our brain and brawn to fight a sterner, longer, more unbending battle against the forces of hell in our own land.

"We return.

"We return from fighting.

"We return fighting. …We will save [Democracy] in the United States of America or know the reason why!"

Bigotry has still not been erased in the United States of America. It takes many forms, from the most subtle discriminatory gesture to blatant, vicious brutality. Hatred born of the irrational fears of those different from oneself smolders deep within our society and occasionally erupts with unpredictability and savagery, like a molten volcano—or a mid-summer riot. Tuskegee Airmen and women returning from World War II found little change in their nation, the country for which they had been fighting.

Though black pioneers in aviation faced vicious bigotry, the successful overcame those barriers. Those brave airmen and airwomen provided a critical legacy for those who would become Tuskegee's Heroes of World War II. In turn, the Tuskegee Airmen created a solid foundation for the future. Black aviators operated in a society that sought to deny opportunity; yet, those that persevered achieved far more than even they had expected or dreamed.

In spite of efforts of U.S. Army Air Forces leaders, the Tuskegee *Experiment* was a success! The Tuskegee *Experience* was a success! Defying all odds, black aviators, who literally flew when opportunity knocked, can be justifiably proud of their contributions. They have led to greater opportunity for all. One has only to read the success stories of these heroes to see that their efforts and their talents have brightened their lives and the military, business, academic, and governmental arenas in which they have excelled. A door to achievement, which was cracked open by black pioneer airmen and women, was thrown wide by those involved with the Tuskegee Experience. Although today's world is far from perfect, those who carry on in the spirit of those heroes can aspire to lofty heights if they, too, persevere.

Blacks can fly, manage, build, and service everything from helicopters and balloons to space shuttles and airports. Cockpits exist in national and international airlines, domestic and transoceanic cargo, and express mail carriers, and careers can be forged in air traffic control and with other governmental and civilian contractors. Positions can be obtained as flight instructors and corporate pilots-in-command and in military aircraft like the ones that brought hero status to Bullard and to the Red Tails, aircraft that continue to attract crewmembers who are the cream of the crop.

Those in military service who pilot and perform crew duties in the nation's modern jets can taste the sweet flavor of promotion just as those who piloted and performed as members of crews during World War II.

top left and bottom left
Daniel "Chappie" James, first black to achieve the rank of general, United States Air Force, flew in three wars. *U.S. Air Force Museum*

bottom right
Major General Lucius Theus entered the U.S. Army Air Corps as a private in 1942. He retired more than 30 years later as a Major General commanding the USAF Accounting and Finance Center, Denver, CO. *Maj. Gen. Lucius Theus, USAF Retired*

Leading the way for today's successful officers were Tuskegee Airmen like Lt. Gen. Benjamin O. Davis, Jr., the first black in the U.S. Air Corps to reach flag rank; Gen. Daniel "Chappie" James, the first to wear the four stars of a full General; and Maj. Gen. Lucius Theus who, in his impressive rise from enlisted man to flag rank, is the first black USAF support officer so promoted. Theus represents his own remarkable achievements as well as the achievements of the thousands of support men and women without whose contribution the Tuskegee Experiment would not have succeeded. It took teamwork!

Theus, currently President of U.S. Associates, consultants in civic affairs, human resources, and business management, enjoyed an auspicious thirty-six year career in the military of his nation. Prior to his current role, he was Director of Civic Affairs for Allied-Signal Corporation and for The Bendix Corporation. He retired from military service in the dual position of Commanding General of the Air Force Accounting and Finance Center and USAF Director of Accounting and Finance with offices in Denver, Colorado, and at the Pentagon, Washington, DC.

In addition to other bases in the contiguous United States, Theus served at Keesler, Tuskegee,

These Are Our Finest,
by John H. Young, III.

Not even the sun was bright that day down in Alabama when a handful of Negro cadets went out to add the highways of the military eagle to the paths of the man of color …went out to answer their Nation's chimerical question: Can a Negro fly and fight in airplanes?

The sun shines right, when things are bright; when the wind blows strong and free.

The challenge was great and the terms were harsh.

Go down there, black boy …way down to Alabama. Sweat out your days and sit out your nights. And show me …black boy …show me you can fly.

The challenge was met by hearts that were strong. From hamlets, towns and cities …from States like Oregon, Maine, California and Georgia …came Jones, Smith, Richards, and Washington; in a long steady procession they came to Tuskegee. The days were long. Hup, hup, hup, ho.

Hold that nose down, Mister. Dammit, get that wing up. Can't you see that horizon, Mister? Quit holding that rudder in turns.

But the journey was swift.

From plane to plane, from light to heavy …from Stearman to Bee Tee to Aye Tee to Thunderbolt …from Tuskegee to Italy.

Could a Negro fly military planes? Well, here they were in Italy and there sat their P-40s cooling on the ramp.

Pretty good, black boy, p-r-e-t-t-y g-o-o-d. But wait a minute. That's not all. Can you FIGHT in planes? Will you come scudding home with the first burst of fire? Better take it easy for a while, black boy …right now, better just fly over yonder and shoot up some trains.

Then Anzio. LSTs rolled up to the beaches and dropped out Americans. The going was rough. Blood everywhere. Enemy planes pounding, strafing. …pushing us back. Suddenly, swiftly and surely …out of nowhere came the friendly whine of P-40s. Down, down, down went Jerry. On the beaches, Americans moved forward and slept peacefully that night.

O tell me, lad, of that day long ago; of the "black boy" from Tuskegee over Anzio.

The rest is a matter of record…a thrill-packed saga of the men who fought in the skies over Europe to save a seemingly ungrateful Nation. But these men are not whole …not full; the best of them lies buried. For were they not given escort by an invincible ghost-like squadron of planes? And where are those who won't come back? Remember "Red" Dawson, "Big" Davis and Sindat-Singh?

"When shall we three meet again?
In thunder, lightning or in rain?"

Heroes? Men are heroes …these are more than men. They are the valiant, the brave …the "black boy" tried and true. They are the exploited, the expendable …the birds with clipped wings. They are the noble, the greatest …these are our finest.

Hats off to the men who tried. Hats off to the men who cried. Hats off …to the men who died.

---

and Lockbourne Army Air Fields; and in Germany, France, Greece, and South Vietnam. In one of his most essential roles, Major General Theus chaired the Inter-Service Task Force whose recommendations led directly to the establishment of a race relations education program and to the current Defense Equal Opportunity Management Institute.

Few realize that Lucius Theus took and successfully passed the exam for pilot training in Chicago in 1942. He said, "My papers got lost in the mill somewhere and, in the meantime, the draft board said it was time to join. At the end of 1942, I took basic training at Keesler Field, Biloxi, Mississippi. Invited by the squadron commander to be permanent party, I went to Atlanta University for a clerical course and was reassigned back to Keesler. Our squadron provided basic training to general inductees and to aviation cadets, some of whom were headed for Tuskegee for pilot training. I was elevated fairly rapidly to First Sergeant and, though they later found my papers that would reassign me to Tuskegee for pilot training, I agreed to remain where I was. At the end of the war, I went to OCS at Maxwell Air Force Base, graduated second in my class, and was assigned to Tuskegee Army Air Field to be Squadron Adjutant in Squadron C, the housekeeping squadron."

Although artist Harry J. Schaare expressed an attitude typical of many pilots when he said, "We used to call non-pilots *Ground Pounders, Paddle Feet, Gravel Agitators, Grunts*," it is undeniably true that for every pilot of an aircraft, there are at

least ten non-rated persons seeing to that aircraft's performance.

General Theus said, "We have not given sufficient credit to non-rated individuals who supported the combat effort of our fighters. Without good ground people, you cannot operate."

He continued, "The support people at Tuskegee were perhaps the most dedicated that I encountered. They knew the spotlight was on the whole operation and that an essential portion had to be accomplished by them. They were determined to provide the very best service that they could to these embryo pilots, knowing that they held the key to whether the whole experiment would be successful or not."

General Theus commended, among others, the staff of the medical group and hospital. He named the engineering department, young men and women deserving nothing but praise. He said, too, "We had a couple of Women's Army Air Corps (WAAC) squadrons. They, too, did absolutely phenomenal work. I may be a bit biased, since I'm non-rated myself, but I think the support people deserve far more credit than they have received.

"Whoever did the assignments to TAAF combed the recruits quite carefully. It was considered to be quite an elite assignment and you had to be good or you didn't get there. I think the most important thing about the Tuskegee experience is that you could see throughout all the ranks an understanding that it had to succeed and that you, as an individual, were not going to hinder that success in any way by not giving your very best. Having a very understanding but tough commander, then-Colonel Davis, added to this. He did not take any baloney whatsoever. There is no question in our minds that with a lesser commander, the success would not have been as great, if successful at all. He understood. He knew what he had riding on his shoulders and that was the success or failure of this total experiment. I have great admiration for the man."

Everyone with a part in the Tuskegee Experiment were willing to make an extra effort toward its success. Recognizing that this was one chance to make a contribution—to the world, to one's race, to oneself—there was an appreciation of the importance of every job. Every person had to do his or her very best.

"Those of us who had the Tuskegee experience," said General Theus, "have always been grateful for it because we found that the discipline and the values that were instilled stood us in great stead throughout the rest of our careers and lives."

A woman who gives credit to her parents for the discipline and values that were instilled in her youth is Doctor Bessie Delany, the second black woman to

course in aviation, provided materials for summer school classes for teachers at Wright-Patterson Air Force Base, Ohio and directed an Aviation Career Exposure, "ACE," program in New York and New Jersey high schools. Her *Fly With Me* coloring book was based on her true story and encouraged children toward lofty aspirations through her contributions to aviation history.

To be black and to be a woman escalated one's difficulties in the "white man's world" of aviation. But, Bessie Coleman, Willa Brown, Ida Van Smith, the Tuskegee nurses, CPTP pilots, WAACs (later, WAC), and many others bore racism stoically.

Lieutenant Colonel Charity Adams Earley joined the WAAC in 1942. She became the first black officer and the commander of the 6888th Central Postal Directory Battalion—the only organization of black women to serve overseas during World War II. She wrote, "The presence of women in the Army was resented by many because, traditionally, the military was male. The resentment was doubled by the service of Negro women because the laws, customs, and mores of the World War II era denigrated and discriminated

obtain a license to practice dentistry in the states of New York and North Carolina. About bigotry, Dr. Delany said, "It was bad enough to be discriminated against by white people because I was colored. But then, my own people would discriminate against me because I was a woman! …if you think it was hard for colored men, honey, colored women were on the bottom. Yes, sir! Colored women took it from all angles!"

Aviatrix Ida Van Smith rose above discrimination. A retired school teacher, Smith produced and hosted a weekly television show in conjunction with her famous Ida Van Smith Flight Clubs, designed to involve young people and to inspire them toward careers in aviation and space. She taught a college

against Negroes. Negro males had been systematically degraded and mistreated in the civilian world, and the presence of successfully performing Negro women on the scene increased their resentment. The efforts of the women to be supportive of the men was mistaken for competition and patronage. We lived with these attitudes with dignity."

Earley continued, "The trailblazing by the women who served in the military during World War II has been virtually ignored and forgotten. …In truth, I have accomplished much since my military service. I have opened a few doors, broken a few barriers, and, I hope, smoothed the way for the next generation. The problems that were my concern dur-

These Are Our Finest

Men of the 99th Fighter Squadron and of the 332nd Fighter Group…

Brave "Black Men" who had to endure the rigors of pilot training and go on to win their wings against all the forces of home-grown bigotry.

They are the brave, the valiant, the lonely eagles. The rest is a matter of record.…

A thrill-packed saga of bold black pilots who fought against Nazism and Fascism in the skies of North Africa and Europe to help save our nation…

These are our finest.

Total number of pilots graduated at Tuskegee: 992.

Awards: Legion of Merit. Silver Star. Purple Heart. Distinguished Flying Cross. Soldiers Medal, Bronze Star. Air Medal and Clusters.

Total killed in action: 66.

World War II—1941-1945. *Roy E. La Grone*

THESE ARE OUR FINEST.
MEN OF THE 99th FIGHTER SQUADRON
AND THE 332nd FIGHTER GROUP…
BRAVE "BLACK MEN," WHO HAD TO
ENDURE THE RIGORS OF PILOT
TRAINING AND GO ON TO WIN
THEIR WINGS AGAINST ALL THE
FORCES OF HOME-GROWN BIGOTRY.
THEY ARE THE BRAVE, THE VALIANT,
THE LONELY EAGLES. THE REST IS
A MATTER OF RECORD… A THRILL-
PACKED SAGA OF BOLD BLACK PILOTS
WHO FOUGHT AGAINST NAZISM AND
FASCISM IN THE SKIES OF NORTH
AFRICA AND EUROPE TO HELP SAVE
OUR NATION…
THESE ARE OUR FINEST.
TOTAL NUMBER OF PILOTS
GRADUATED AT TUSKEGEE 992.
AWARDS: LEGION OF MERIT. SILVER STAR.
PURPLE HEART. DISTINGUISHED FLYING
CROSS. SOLDIER MEDAL, BRONZE STAR,
AIR MEDAL AND CLUSTERS.
TOTAL KILLED IN ACTION 66
WORLD WAR II 1941-1945

ROY E. LA GRONE
TUSKEGEE AIRMEN. S.L. 79

332ND FIGHTER GROUP

William E. Brown, who later became a Major General, with his F-86 Sabre. Brown completed 225 combat missions in the Korean and Vietnam wars. *U.S. Air Force Museum*

ing my service, and to which I have devoted my energies, are still with us—but I keep trying."

Without the courage and commitment of pioneering women in military aviation, modern pilots and officers like Stayce Harris, Marcella A. Hayes, and Theresa Clayborn would not have found their roles possible. Harris has piloted a C-141 for the 14th Military Airlift Squadron. Hayes is the first black woman to complete pilot training in the U.S. Army, an Army Helicopter pilot who earned her wings in 1979. Clayborn in 1984 became the first black woman to complete jet pilot training in the U.S. Air Force.

General Colin L. Powell was the first black chairman of the Joint Chiefs of Staff, attaining the highest ranking of a black military officer in U.S. history. Black women, too, have achieved a status in U.S. military service that was previously beyond their wildest dreams! The Army's Hazel Johnson-Brown, Chief of the Army Nurse Corps; Air Force's Marcelite Harris, a Vice Commander of the Oklahoma City Air Logistics Center; and Air National Guard's Irene Trowell-Harris, Nursing Assistant to the Director of Nursing, Office of the Inspector General, Air Force Surgeon General, can vouch for that. All three were promoted to Brigadier General.

Gene Carter, Lou Purnell, Mildred "Mike" Carter, and Lt. Gen. Benjamin O. Davis, Jr., photographed at a Tuskegee Airmen convention. *Lt. Col. Gene Carter, USAF Retired*

Brigadier General Irene Trowell-Harris, first black female general in the 357-year history of the Army and Air National Guards. General Trowell-Harris was also the first nurse to command an Air National Guard clinic. *Brig. Gen. Irene Trowell-Harris*

General Trowell-Harris, who labored for hours with her ten brothers and sisters in cotton fields as a child in rural South Carolina, holds a Master's degree from Yale University and a Doctorate of Philosophy in health education from Columbia University. She is the first black female general in the 357-year history of the Army and Air Force National Guards and the third female general in the military's oldest force. On the way to winning those stars, General Trowell-Harris was the first nurse in Air National Guard history to command a medical clinic.

A tireless worker, Trowell-Harris credits her mother with encouragement that challenged her to achieve. She said, "My mother always said, 'Stay in school. Stay in church.'" When confronted with inevitable racism, Trowell-Harris said, "I saw no point in getting defensive. I stayed focused."

As a contemporary officer, General Trowell-Harris recalls the past as she looks to the future. She said, "People helped me and I want to help others. I want to encourage young people to work hard, to be honest, and to stay away from drugs and alcohol—to be model citizens."

Another contemporary, Maj. Gen. Russell C. Davis, is the Vice Chief National Guard Bureau, Wahington DC. Davis could not help but have been inspired by the Tuskegee Airmen, as he was born in Tuskegee, Alabama, in 1938, and attended Tuskegee Institute High School and Tuskegee University. In pursuit of excellence, Russ Davis, a command pilot with more than 4,700 flying hours, earned a juris doctor degree and, in addition to his military service to our nation, practiced law from 1969 to 1978. He

served in numerous positions of increased authority and responsibility including fighter interceptor pilot, flight commander, air operations staff officer, and director of operations, Headquarters, Iowa Air National Guard. In 1991 he assumed his present position. Among other decorations, General Davis has been awarded the Legion of Merit, USAF Meritorious Service Medal, USAF Commendation Medal, Combat Readiness Medal, and the Distinguished Service Medal. A news item noted that Davis was the pilot who delivered the F-4C aircraft flown by the late Chappie James from Washington, DC, to Maxwell Air Force Base, Alabama, for permanent display on the grounds of the General Daniel Chappie James Center, Tuskegee.

Perhaps the Tuskegee Airman who best personifies a link—past, present, and future—is Col. William "Wild Bill" Campbell, USAF (Ret). In 1990, Campbell was honored by the Commonwealth Club of San Francisco as one of a distinguished group of fighter pilots and a news item stated, "Among others who have been compared to 'the knights of old,' fighter pilots are a proud lot—with egos to match their accomplishments as the 'hottest sticks in the world.'" In 1994, Maxwell Air Force Base's Air Command and Staff College (ACSC) hosted its annual Gathering of Eagles, bringing together outstanding individuals in aviation. One of the notable

which the United States was involved, those that were honored by selection for the prestigious aerial demonstration teams like the *Thunderbirds*, and those whose dreams took them to test piloting and outer space.

The achievements of the late Col. John Whitehead, the Air Force's first black experimental test pilot, honor all black aviators who have served as test pilots. During "Mr. Death's" thirty-year career, he logged over 9,500 flying hours with over 5,000 hours in jet aircraft. He flew forty different types of aircraft including fighters, bombers, and transports. Whitehead, a past president of Tuskegee Airmen, Inc., said, "…we honor the leadership, advocacy, and commitment that strengthened the fabric of America by mandating that its Armed Forces be 'color-blind.' Truman's Presidential Order may very well have spared our nation from the disaster of sustained disunity and eventual self-destruction."

Major General Russell C. Davis, then Commanding General, District of Columbia National Guard and now Vice Chief, National Guard Bureau. *Maj. Gen. Russell C. Davis*

Guion "Guy" Bluford, the first black American to be selected for astronaut training and the first to fly in space. *Official NASA Photograph, U.S. Air Force Museum*

aviators was Campbell—the pilot credited with flying the 99th's first combat mission. The son of the late Thomas Campbell, the first black Extension Agent in U.S. Governmental Agricultural Service, "Wild Bill" is the brother of the 6888th Battalion's Abbie Noel Campbell, Executive Officer, second-in-command to Lieutenant Colonel Earley in the only organization of black women to serve overseas during World War II and of Thomas Campbell, Jr., a medical officer with the same organization. Bill Campbell is the proud father of three sons, one of whom, Steve, carried on the tradition of the fighter pilots of the 99th by having flown in the USAF as an F-15 pilot. La Grone's art work captured the link of father and son, past and future.

Four members of the 332nd Fighter Group in La Grone's painting, "Jet Instructors at the Fighter School 1949," can be interpreted to represent all black aviators who became jet pilots. Those at the leading edge of mandated integration of the Air Force, four former Tuskegee Airmen—Maj. Lewis "Lew" Lynch, and Colonels Vernon Vincent "V.V." Haywood, Henry "Herky" Perry and John "Mr. Death" Whitehead—served in the Jet Pilot Training Instructor Program, the 3525th Pilot Training Wing (PTW), Williams Air Force Base, Arizona. La Grone's paintings have represented the many by spotlighting those who served in three wars in

Red-Tailed Angels
by Damon Chandler

As a young black boy in America,
I dreamed of World War II.
How sad that I had been born too late,
and there was nothing for me to do.
So I watched every movie that was ever made,
about killing the "Nazis" and "Japs."
When Americans died, I sat there and cried,
as the bugler sadly blew taps.

But one thing never phased me,
as I gazed upon Hollywood's wars.
That blacks were never included,
in the fighting beyond our shores.
John Wayne destroyed whole armies,
even Tarzan swung on the scene.
We blacks were safe on the homefront,
our job was to cook and clean.

But reality was Pearl Harbor,
December seventh, nineteen forty-one.
And a young black seaman who bravely
faced death and didn't run.
I had never heard of Dorie Miller,
who manned a blazing gun.
Four Zeros passed within his sights,
and went to their rising sun.

Nor had I heard of red-tailed angels,
black warriors who graced the skies.
Their Mustangs streaked the heavens,
And Germans fell like flies.
Escorting bombers to their targets,
these angels performed so well.
German fighters never downed a bomber,
as black angels sent them to hell.

The German high-command took notice,
and kept track of group Three Thirty-Two.
They had heard black men were fighting,
and had seen what Jesse Owens could do.
They also remembered a previous war,
and what a soldier named Johnson had done.
Twenty-five Germans slipped into his trench,
only three could get out and run.

But America had yet to acknowledge
the courage and skill of her own.
Save the bomber crew who knew the truth
who the red-tailed angels brought home.
These brave black men of yesteryear,
four hundred and fifty strong,
only sixty-six would lose their lives,
and white American was proven wrong.

This highly motivated unit,
had fought two separate wars.
In the air, they fought the Germans,
on the ground, a different cause.
While they fought for freedom for others
things remained as they had been.
Black America continued to suffer,
because of difference in the color of skin.

But they endured and fought the good fight,
black warriors of the sky.
Americans denied equal citizenship,
were more than willing to die.
To help cut away the cancerous growth of hate,
to help change the racist view.
These valiant men were given the chance
to show what they could do.

Over forty years after they took to the air,
there are some who still live today.
Much older, there's stil a twinkle in their eyes,
though the hair, if any, is gray.
They blazed a trail of glory,
red-tailed angels, wings of gold.
And now we know the true story,
of men, courageous and bold.

Looking back, how sad to think the truth,
for so long was terribly denied.
As a young black man in America,
my soul was filled with pride.
That we too had faced the enemies,
their evil dreams were dead.
And perhaps someone would take notice,
that our blood was also red.
-from the TAI Newslettler

Preflight navigation. *U.S. Air Force Museum*

Enlisted Crew Chiefs changing a propeller governor in a muddy field. *U.S. Air Force Museum*

Lieutenant Colonel Charity Earley with the 6888th Central Postal Directory's two Captain Campbells: Capt. Thomas Campbell, medical officer, and his sister, Capt. Abbie Noel Campbell, executive officer. Both Campbells are brother and sister, respectively, to the 99th Fighter Squadron's outstanding pilot, William Campbell, Tuskegee Airman. *Lt. Col. Charity Adams Earley, U.S. Army, Retired*

Cockpit Check, Randolph Field

From the U.S. Air Force Art Program. *Roy E. La Grone*

Two Comrades In Arms

From the U.S. Air Force Art Program. *Roy E. La Grone*

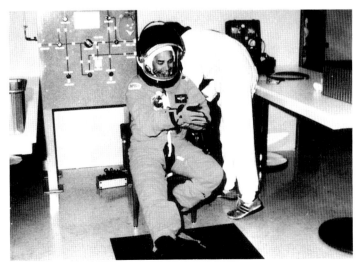

Fred Gregory is suited up. *Official NASA photo, U.S. Air Force Museum*

Official NASA photo of Fred Gregory, first black American to command a Shuttle mission. *U.S. Air Force Museum*

Roy E. La Grone flanked by Lee "Buddy" Archer, on his right, and Dr. Roscoe C. Brown, Jr., at one of the artist's favorite sites, the Society of Illustrators, New York. *Pat Whyte*

The sky holds no limits for those who are willing to work and who capitalize on opportunity when it appears. As Lt. Cdr. Donnie Cochran was the first African American member of the U.S. Navy flight demonstration squadron, the *Blue Angels,* La Grone's painting of Maj. Gen. Lloyd W. "Fig" Newton shows him as the first black to perform as a pilot with the USAF Aerial Demonstration Team, the *Thunderbirds.* It doesn't show that Newton held several positions with the elite team, serving as narrator and advance coordinator before piloting with the elite group. La Grone's painting of four black male astronauts, Colonels Guion S. Bluford, Jr., Charles Bolden, Frederick D. Gregory, and the astronaut who died January 1986 in the tragic *Challenger* disaster, the late Ronald E. McNair, is nothing if it is not an inspiration! It represents the achievement of the highest in ideals and dreams. It does not show that black women can aspire to the same dizzying heights, although Dr. Mae Jemison, as the first U.S. black woman in NASA's astronaut program, fulfilled that role. Born in 1956, she was in her thirties in September 1992 when, as a mission specialist, she blasted into outer space (and history books) aboard the space shuttle *Endeavor.* Jemison, only sixteen when she matriculated, obtained her bachelor's

degree in chemical engineering in 1977 from Stanford University, California, and her medical degree from Cornell University four years later. Also a student of African history and African American studies, Jemison is multilingual, speaking Russian, Swahili, and Japanese in addition to her native English. From 1983 to 1985, she served as a Peace Corps medical officer managing a health care delivery system in Sierra Leone and Liberia, Africa.

Tuskegee's airmen and airwomen established a legacy for excellence. They gave of themselves to the best of their ability and earned their title of HEROES. As Tuskegee Airman Lee Archer indicated in the following speech, that legacy can be a plea:

"The Negro [sic] who hangs around the corners of the streets or lives in Grog-shops, or by gambling, or who has no higher ambitions to serve, is by his vocation forging fetters for the slave, and is to all interests and purposes a curse to his race. Black men and women must adhere to a number of moral imperatives if we are to raise ourselves—the primary imperative being, 'Each one in whatever his place, determined to do what he can for himself and the race,' and the second imperative, 'We must enter into an era of self-improvement in which skilled, talented, educated, and creative black men and women strive to increase our numbers and our cohesiveness.'

Lieutenant Colonel Charity A. Earley, who was on the rifle team at Wilberforce University, Ohio, under Brigadier General B. O. Davis, Sr. She rose to command the 6888th Central Postal Directory, Headquarters, European Theater of Operations (ETO) during World War II. *Lt. Col. Charity Adams Earley, U.S. Army, Retired*

Debriefing after the mission was as important as preflight planning. Shown here are Lts. Howard Baugh and Richard C. Bolling, 99th Fighter Squadron. On 10 July 1943, Bolling was shot down while patrolling the Sicilian invasion area; he parachuted from his burning fighter and was picked up at sea by an American destroyer. *Lt. Col. Gene Carter, USAF Retired*

The Joshua Tree and
The F-15 Eagle

From the U.S. Air Force Art Program. *Roy E. La Grone*

Operation Provide Hope

La Grone was impressed with the gray skies of Siberia. From the U.S. Air Force Art Program. *Roy E. La Grone*

"I am not suggesting we attempt to build a race of super-blacks. What I am suggesting is depth in leadership. It cannot be denied that in every nation or race leadership devolves to a small fraction of the group (usually about 10 percent) and leadership continuity is essentially maintained by that group.

"This is what the Tuskegee Airmen were all about and what it should be about in the future. ...To quote John Rock..., 'We cannot expect to occupy a much better position than we do now until we shall have educated and wealthy men and women who can wield power that cannot be misunderstood. Then, and not till then, will the tongue of slander be silenced and the lip of prejudice sealed.' I recommend to the Tuskegee Airmen Organization a vigilant search and sponsorship of the young who will replace us."

This collection of paintings—the invaluable contribution of the late Roy La Grone—capture what "the Tuskegee Airmen were all about." For a nation and world badly in need of true heroes, the unique men and women represented in his art can claim that title. They can be remembered with admiration and remain as inspiration for today and for tomorrow.

opposite
Mechanics of the 332nd Fighter Group, 15th Air Force, repair the engine of a P-51 Mustang: Three of them are Staff Sergeants Calvin P. Thierry, William E. Pitts, and Harold T. Cobb. *Official USAF Photo, Roy E. La Grone*

Prisoners of World War II

Lieutenant Colonel Alexander Jefferson, USAF Retired, graduated from single engine pilot training at Tuskcgcc Army Air Field in January 1944. After three months of Combat Training at Selfridge Field, Michigan, he joined the 301st Fighter Squadron, 332nd Fighter Group in Ramitelli, Italy. Jefferson's combat flying came to an abrupt halt when he was shot out of the sky by Nazi guns. Downed and captured on the same strafing mission as Robert Daniels, he admitted, "It made me feel better to have some one I knew with me…misery loves company."

Jefferson's experiences are representative of those with whom he served and with whom he was incarcerated. He said, "Assigned as a P-51 fighter pilot with the 332nd, I flew 18 long range missions escorting B-17 and B-24 bombers. On August 12, 1944, three days prior to the invasion of Southern France, I was shot down by ground fire while strafing radar stations on the coast. Captured by German troops, I was interned for nine months as a prisoner of war."

In *Tony Brown's Journal*, Jefferson wrote, "They put is in solitary for approximately three days. They led me out to an officer and a German officer said, 'Have a seat, Lieutenant.' …in front of him was a large book and on the front of it, it says, 'Negroes. 332nd Fighter Group. Red Tails.' …The S.S. tried to take us over. Nothing happened and, to tell you the truth, we weren't afraid."1

Jefferson spent five months in Stalag Luft III, 80 miles east of Berlin. When the Russian offensive began in January of 1945, the Nazis transferred American, British and French prisoners to Stalag Luft VII-A, Mooseburg, approximately 20 miles north of Dachau. He said, "The first 80 kilometers were a forced march in temperatures of 20- to 30-degrees below zero. Prisoners were then put in the famous '40s and 80s,' [boxcars] with 70 to 80 men to a car. After liberation by Patton's Third Army and prior to returning home, I visited Dachau to witness the results of the atrocities committed by the Nazis."

He witnessed tragedy in German youth. He said, "The most frightening thing that ever happened to me was when I saw Hitler youth. When they saw us, they started their tirade about 'ya, ya, ya,' saluting and heel clicking. It reminded me of the Ku Klux Klan, the most horrible thoughts out of all my experience. That's the only time I really became frightened. …The terror of thinking how children's minds can be so warped."2

Returned to Tuskegee AAF, Jefferson became an instrument instructor in advanced training, was discharged from active duty in 1947, and retired from the reserves in 1969 after having served as Staff Operations and Training Officer of the 9505th Air Recovery Squadron. Highly educated, he became an elementary school science teacher in Detroit, Michigan and retired as assistant principal in 1979. An active member of various organizations, Jefferson serves on a voluntary basis as an admissions counselor for the U.S. Air Force Academy and Air Force ROTC.

In chronicling the heroism of the Tuskegee Airmen, we must remember those that faced the added horror of being held as Prisoners of War! Thanks to the efforts of the Red Cross, they were minimally clothed and fed, but the hardship and deprivation of prison life was exacerbated by the constant anxiety of not knowing what would become of them.

The following list of 32 Black American Prisoners of the Second World War was compiled by Colonel Jefferson, who supplied the details to the best of his knowledge. All were truly *Tuskegee's Heroes!*

Name:	Camp:	Shot Down Over:
Bolden, Edgar L.	Stalag I (Barth)	
Brown, Harold, H.	Stalag I	Linz, Austria
Brown, Gene	Stalag I	
Brantley, Charles V.	Stalag Luft VII A (Mooseburg)	
Daniels, Robert H.	Stalag Luft III (Sagan), VII A	Southern France
Driver, Clarence N.	Stalag Luft VII A	Danube River, strafing
Gaines, Thurston. L., Jr.	Stalag Luft VII A	Train Station, strafing
Gaither, Roger	Stalag Luft III, VII A	
Golden, Newman	Undocumented	
Gorham, Alfred	Undocumented	
Gould, Cornelius	Stalag I	
Griffin, William	Stalag I	
Hathcock, Lloyd S.	Stalag Luft III, VII A	
Hudson, Lincoln T.	Stalag Luft III A (Nuremberg), VII A	
Iles, George	Stalag Luft III A, VII A Near Swiss Border	
Jefferson, Alexander	Stalag Luft III, VII A	Southern France
Johnson, Langston	Undocumented	
Lewis, Joe	Stalag Luft VII A	
Long, Wilbur F.	Stalag Luft VII A	Blechhammer, Poland
Macon, Richard	Stalag Luft III A, VII A	Southern France
McCreary, Walter	Stalag Luft III, VII A	
McDaniel, Armour	Stalag Luft VII A	Berlin, Germany
Morgan, Woodrow	Stalag Luft III	
Penn, Starling	Stalag I	Linz, Austria
Smith, Lewis C.	Stalag Luft III	
Smith, Luther	Hospital	
Thompson, Floyd	Stalag Luft VII A	
White, Hugh	Stalag Luft VII A	
Williams, Charles T.	Stalag Luft VII A	
Williams, Kenneth I.	Stalag Luft VII A	
Wise, Henry	Stalag I	
Woods, Carrol S.	Stalag Luft VII A	

After having his aircraft shot down by ground fire, then-Lieutenant Jefferson was captured by German troops and interned for nine months as a prisoner of war. He was later liberated by Patton's Third Army. *Lt. Col. Alexander Jefferson, USAF Retired*

Aerial Victories by Tuskegee Airmen in World War II

Date	Name	Squadron	A/C Flown	A/C Destroyed	Number
2 Jul 43	1/Lt Charles B. Hall	99th	P-40L	FW-190	1.0
27 Jan 44	2/Lt Clarence W. Allen	99th	P-40L	FW-190	0.5
	1/Lt Willie Ashley, Jr.	99th	P-40L	FW-190	1.0
	2/Lt Charles P. Bailey	99th	P-40L	FW-190	1.0
	1/Lt Howard L. Baugh	99th	P-40L	FW-190	1.0
	Capt. Lemuel R. Custis	99th	P-40L	FW-190	1.0
	1/Lt Robert W. Deiz	99th	P-40L	FW-190	2.0
	2/Lt Wilson V. Eagleson	99th	P-40L	FW-190	1.0
	1/Lt Leaon C. Roberts	99th	P-40L	FW-190	1.0
	2/Lt Lewis C. Smith	99th	P-40L	FW-190	1.0
	1/Lt Edward L. Toppins	99th	P-40L	FW-190	1.0
28 Jan 44	Capt. Charles B. Hall	99th	P-40L	FW-190	1.0
	Capt. Charles B. Hall	99th	P-40L	Me-109	1.0
5 Feb 44	1/Lt Elwood T. Driver	99th	P-40L	FW-190	1.0
7 Feb 44	2/Lt Wilson V. Eagleson	99th	P-40L	FW-190	1.0
	2/Lt Leonard M. Jackson	99th	P-40L	FW-190	1.0
	1/Lt Clinton B. Mills	99th	P-40L	FW-190	1.0
9 Jun 44	1/Lt Charles M. Bussey	302nd	P-47	Me-109	1.0
	2/Lt Frederick Funderburg	301st	P-47	Me-109	2.0
	1/Lt Melvin T. Jackson	302nd	P-47	Me-109	1.0
	1/Lt Wendell O. Pruitt	302nd	P-47	Me-109	1.0
12 Jul 44	Capt. Joseph D. Elsberry	301st	P-51	FW-190	3.0
	1/Lt Harold E. Sawyer	301st	P-51	FW-190	1.0
16 Jul 44	1/Lt Alfonzo W. Davis	99th	P-51	Mc. 205	1.0
	2/Lt William W. Green, Jr.	302nd	P-51	Mc. 205	1.0
17 Jul 44	1/Lt Luther H. Smith, Jr.	302nd	P-51	Me-109	1.0
	2/Lt Robert H. Smith	302nd	P-51	Me-109	1.0
	1/Lt Laurence D. Wilkens	302nd	P-51	Me-109	1.0
18 Jul 44	2/Lt Lee A. Archer	302nd	P-51	Me-109	1.0
	1/Lt Charles P. Bailey	99th	P-51	FW-190	1.0
	1/Lt Weldon K. Groves	302nd	P-51	Me-109	1.0

Date	Name	Squadron	A/C Flown	A/C Destroyed	Number
	1/Lt Jack D. Holsclaw	100th	P-51	Me-109	2.0
	2/Lt Clarence D. Lester	100th	P-51	Me-109	3.0
	2/Lt Roger Romine	302nd	P-51	Me-109	1.0
	Capt. Edward Toppins	99th	P-51	FW-190	1.0
	2/Lt Hugh S. Warner	302nd	P-51	Me-109	1.0
20 July 44	Capt. Joseph D. Elsberry	301st	P-51	Me-109	1.0
	1/Lt Langdon E. Johnson	100th	P-51	Me-109	1.0
	Capt. Armour G. McDanaiel	301st	P-51	Me-109	1.0
	2/Lt Walter J.A. Palmer	100th	P-51	Me-109	1.0
	Capt. Edward L. Toppins	99th	P-51	FW-190	1.0
25 Jul 44	1/Lt Harold E. Sawyer	301st	P-51	Me-109	1.0
26 Jul 44	1/Lt Freddie F. Hutchins	302nd	P-51	Me-109	1.0
	1/Lt Leonard M. Jackson	99th	P-51	Me-109	1.0
	2/Lt Roger Romine	302nd	P-51	Me-109	1.0
	Capt. Edward L. Toppins	99th	P-51	Me-109	1.0
27 Jul 44	1/Lt Edward C. Gleed	301st	P-51	FW-190	2.0
	2/Lt Alfred M. Gorham	301st	P-51	FW-190	2.0
	Capt. Claude B. Govan	301st	P-51	Me-109	1.0
	2/Lt Richard W. Hall	100th	P-51	Me-109	1.0
	1/Lt Leonard M. Jackson	99th	P-51	Me-109	1.0
	1/lt Felix J. KIrkpatrick	302nd	P-51	Me-109	1.0
30 Jul 44	2/Lt Carl E. Johnson	100th	P-51	Re.2001	1.0
14 Aug 44	2/Lt George M. Rhodes, Jr.	100th	P-51	FW-190	1.0
23 Aug 44	F/O William L. Hill	302nd	P-51	Me-109	1.0
24 Aug 44	1/Lt John F. Briggs	100th	P-51	Me-109	1.0
	1/Lt Charles E. McGee	302nd	P-51	FW-190	1.0
	1/Lt William H. Thomas	302nd	P-51	FW-190	1.0
12 Oct 44	1/Lt Lee A. Archer	302nd	P-51	Me-109	3.0
	Capt. Milton R. Brooks	302nd	P-51	Me-109	1.0
	1/Lt William W. Green, Jr.	302nd	P-51	He-111	1.0
	Capt. Wendell O. Pruitt	302nd	P-51	He-111	1.0

Date	Name	Squadron	A/C Flown	A/C Destroyed	Number
	Capt. Wendell O. Pruitt	302nd	P-51	Me-109	1.0
	1/Lt Roger Romine	302nd	P-51	Me-109	1.0
	1/Lt Luther H. Smith, Jr.	302nd	P-51	He-111	1.0
16 Nov 44	Capt. Luke J. Weathers, Jr.	302nd	P-51	Me-109	2.0
24 Mar 45	2/Lt Charles V. Brantley	100th	P-51	Me-262	1.0
	1/Lt Roscoe C. Brown	100th	P-51	Me-262	1.0
	1/Lt Earl R. Lane	100th	P-51	Me-262	1.0
31 Mar 45	2/Lt Raul W. Bell	100th	P-51	FW-190	1.0
	2/Lt Thomas P. Braswell	99th	P-51	FW-190	1.0
	1/Lt Roscoe C. Brown	100th	P-51	FW-190	1.0
	Maj. William A. Campbell	99th	P-51	Me-109	1.0
	2/Lt John W. Davis	99th	P-51	Me-109	1.0
	2/Lt James L. Hall	99th	P-51	Me-109	1.0
	1/Lt Earl R, Lane	100th	P-51	Me-109	1.0
	F/O John H. Lyle	100th	P-51	Me-109	1.0
	1/Lt Daniel L. Rich	99th	P-51	Mc-109	1.0
	2/Lt Hugh J. White	99th	P-51	Me-109	1.0
	1/Lt Robert W. Williams	100th	P-51	FW-190	2.0
	2/Lt Bertram W. Wilson, Jr.	100th	P-51	FW-190	1.0
1 Apr 45	2/Lt Carl E. Carey	301st	P-51	FW-190	2.0
	2/Lt John E. Edwards	301st	P-51	Me-109	2.0
	F/O James H. Fischer	301st	P-51	FW-190	1.0
	2/Lt Walter P. Maanning	301st	P-51	FW-190	1.0
	2/Lt Harold M. Morris	301st	P-51	FW-190	1.0
	1/Lt Harry T. Stewart	301st	P-51	FW-190	3.0
	1/Lt Charles L. White	301st	P-51	Me-109	2.0
15 Apr 45	1/Lt Jimmy Lanham	301st	P-51	Me-109	1.0
16 Apr 45	1/Lt William S. Price III	301st	P-51	Me-109	1.0
26 Apr 45	2/Lt Thomas W. Jefferson	301st	P-51	Me-109	2.0
	1/Lt Jimmy Lanham	301st	P-51	Me-109	1.0
	2/Lt Richard A. Simons	100th	P-51	Me-109	1.0

OFFICIAL TOTALS:

99th FIGHTER SQUADRON	30.5
100th FIGHTER SQUADRON	22.0
301st FIGHTER SQUADRON	31.0
302ndFIGHTER SQUADRON	28.0
TOTAL	111.5

Tuskegee Honor Roll

The Following Died on Active Duty between 1941 and 1951:

LET US NOT FORGET

William P. Armstrong
Kenneth Austin
George A. Bates
Lawrence Baissoan
Richard H. Bell
Celsus E. Bequesse
Samuel A. Black
Linson Blackney
Fred L. Brewer
Sidney Brooks
James E. Brothers
Donald E. Brown
James B. Brown
Roger B. Brown
Samuel Bruce
James A. Calhoun
John H. Chavis
Arnold Cisco
George Cisco
James Coleman, Jr.
Coleman Conley
Harry Jay Daniels
John Daniels
Luther R. Davenport
Alfonso Wesley Davis
Richard Davis
Richard A. Dawson
Charles Warren Dickerson
Othel Dickson
Edward Dozier
Alwayne Dunlap

Jerome Edwards
Spurgeon Ellington
Maurice V. Esters
William J. Faulkner
Samuel J. Foreman
Frederick D. Funderburg
Howard C. Gamble
Morris E. Gant
Clemenceau Givings
Walter S. Gladden
Joseph E. Gordon
Robert A. Gordon
George E. Gray
William E. Griffin
Milton R. Hall
Richard W. Hall
Maceo Harris
Thomas L. Hawkins
George Kenneth Hayes
Earl Highbaugh
William E. Hill
Nathaniel M. Hill
Wendell W. Hockaday
Tommy Hood
Stephen Hotesee
Sylvester H. Hurd, Jr.
Oscar D. Hutton, Jr.
Wellington G. Irving
Spencer P. Isabelle
Samuel Jefferson
Charles B. Johnson

Langdon E. Johnson
Edgar Jones
Robert M. Johnson
Horace E. Joseph
Oscar Kenny
Earl Eugene King
Edward Laird
Allen Lane
Carrol N. Langston, Jr.
Erwin Lawrence
Walter I. Lawson
Samuel Leftenant
Wayne Leggins
Albert Manning
Walter P. Manning
Andrew Maples, Jr.
Andrew D. Marshall
Otis E. Marshall
Harold Martin
Vincent Jay Mason
William T. Mattison
Cornelieus May
George McCrumby
James McMullin
Raymond C. McEwen
Faythe A. McGinnis
Paul Mitchell
Frank H. Moody
Roland W. Moody
John Morgan
Sidney Mosely

Neal Nelson
Elton H. Nightengale
Raymond F. Noches
Leland H. Pennington
Francis B. Peoples
Harvey N. Pinkney
James Polkinghorne
Henry Pollard
Driscoll Ponder
John H. Prowell, Jr.
Wendell O. Pruitt
Glen W. Pulliam
Leon Purchase
James C. Ramsey
Nathaniel Rayburg
Ronald Reeves
Emory Robbins
Leon Roberts
Robert C. Robinson
Cornelius Rogers
Roger Romine
Mac Ross
Pearlee E. Saunders
Paul C. Simmons
Alfonso Simmons
Wilmeth Sidat-Singh
John S. Sloan
Graham Smith
Reginald V. Smith
Arnett Stark, Jr.
Charles W. Stephens

Nathaniel C. Stewart
Ross Stewart, Jr.
Roosevelt Stigger
Norvell Stoudmire
Thomas C. Street
John W. Squires
Elmer Taylor
Edward M. Thomas
Edward N. Thompson
Cleodis V. Todd
Edward Toppins
Robert Tresville
Andrew Turner
William Walker
Dudley Watson
Johnson C. Wells
Judson West
Walter Westmoreland
Sherman White
Leonard R. Willette
Eli B. Williams
Leroi Williams
William F. Williams, Jr.
Robert H. Wiggins
Carl J. Woods
Frank N. Wright
James W. Wright
Beryl Wyatt
Albert L. Young

Bibliography

AIR FORCE Magazine, January, 1983.

Allen, Henry; "To Fly, to Brave the Wind, Cornelius Coffey's Dream of Wings," *The Washington Post*, September 26, 1979.

Anderson, C. Alfred "Chief;" *Chief Tells His Own Story* , April 4, 1976.
Black Airmen 1941-1945, Yearbook, Compiled by the Tuskegee Airmen, Inc.

Archer, Lee A., *Tuskegee Airmen Newsletter*. "This is what Tuskegee Airmen was all about and what it should be about in the future." Excerpts from Awards & Recognition Dinner Address, Ft. Carson, Colorado, March 6, 1988.

Brown, Tony,*Tony Brown's Journal*, The Story of America's Black Air Force, Tony Brown Productions, New York, 1983.

Bunch, Lonnie; NASM Education Division, "Pioneers—The 99th Fighter Squadron," *Air and Space*, National Air & Space Museum, Smithsonian Institution Press, January/February 1979.

Carisella, P.J. and Ryan, James W.; *The Black Swallow of Death*, Marlborough House, Inc., Boston, Massachusetts; 1972.

Chandler, Damon, *Red-Tailed Angels*, a Poem; Tuskegee Airmen Newsletter, Courtesy United States Air Force Museum.

Ceasor. Ebraska Dalton; *Mae C. Jemison, First Black Female Astronaut*, New Day Press, Inc., 1992.

Davis, Benjamin O., Jr.; *An Autobiography*, Smithsonian Institution Press, Washington and London, 1991.

Delany, Sarah and A. Elizabeth with A.H.Hearth; *Having Our Say, The Delany Sisters' First 100 Years*; Kodansha America, New York, 1993.

Earley, Charity Adams, *One Woman's Army*, Texas A & M University Press, Military History Series, College Station, Texas, 1989.

Ebony Heroes of Flight, Video aired by the University of Alabama Television Services, July 11, 1985.

Farrar, Fred; "Black Pilots Group, Earned Their Wings the Hard Way," *FAA World*, October, 1976.

Francis, Charles E.; *The Tuskegee Airmen, The Men Who Changed A Nation*, Branden Publishing Company, Boston, MA, 1988.

Gropman, Alan L., Colonel USAF, "Against All Foes," *AIR MAN*, September, 1985.

Gropman, Alan L., *The Air Force Integrates, 1945-1964*, Washington, D.C., 1978.

Hardesty, Von and Pisano, Dominick; *Black Wings*, The American Black in Aviation; National Air & Space Museum, Smithsonian Institution Press, Washington, DC, 1983.

Harris, TSgt Lorenzo D.; "The 'Blue' in Bluford: From Fighter Pilot to Astronaut." *AIRMAN, Official Magazine of the USAF*, February 1984.

Hasdorff, Dr. James C.; Interview with Mr. C. Alfred "Chief" Anderson; United States Air Force Oral History Program; Tuskegee, Alabama, 8-9 June 1981.

Hasdorff, Dr. James C.; Interview with Brigadier General Noel F. Parrish; United States Air Force Oral History Program; San Antonio, Texas, June, 1974.

Hunt, Rufus; Chicago's Air Route Traffic Control Center, FAA. *The Cofey Intersection*.

Hunt, Rufus; *Cornelius R. Coffey, pioneer aviator, mechanic and instructor*, Courtesy, United States Air Force Museum, Dayton, OH.

Letter, History of the 332nd Fighter Group, Oscoda AF, Oscoda, MI, 2 June 1943.

Loving, Neal V., *Loving's Love*, Smithsonian Institution Press, Washington and London, 1994.

Malthaner, T/Sgt. John; *ON GUARD*, National Guard Bureau, Washington, DC, December 1993.

McGovern, James R.; *Black Eagle, General Daniel 'Chappie' James, Jr.*, University of Alabama Press, University, AL, 1985.

Meltzer, Milton; *The Black Americans, A History in Their Own Words 1619-1983*, Harper Trophy, 1984.

Mitchell, Mitch; "FIRST BLOOD, The Saga of the 99th Fighter Squadron," *Air Classics*, February, 1987.

Mockler, Don R., General Editor; *AERONAUTICS, An Authoritative Work Dealing with the Theory and*

Practice of Flying; Issues 1 and 2; National Aeronautics Council, Inc., New York, New York; September, 1940.

Moolman, Valery; "Black Angels, Tuskegee's Awesome Aviators," *The Compass*, Volume LXIII, Number 2, Brooklyn, New York, 1993.

Mullen, Robert W.; *Blacks in America's Wars*, Pathfinder, New York, NY, 1973.

O'Neil, Paul and Editors; *Barnstormers and Speed Kings*, Time-Life Books, Alexandria, Virginia; 1981.

Osur, Alan M.; *Blacks in the Army Air Forces during WWII: The Problem of Race Relations;* Office of Air Force History, Washington, DC, 1977.

OX5 News, "CORNELIUS R. COFFEY." Published by OX5 Aviation Pioneers, Pittsburgh, PA 15216, Volume 35, Number 4, August 1993.

Paszek, Lawrence J., Senior Editor, Office of Air Force History; "SEPARATE BUT EQUAL? The Story of the 99th Fighter Squadron, *Aerospace Historian*, September, 1977.

Purnell, Louis R.; *The Flight of the Bumblebee*, Air & Space Magazine, Smithsonian, October/November, 1989.

Reed, Sam, SSgt., Wing Public Affairs Office. "This Man Has a Story to Tell," *Maxwell-Gunter Dispatch*, February 25, 1994.

Rich, Doris; *Queen Bess, Daredevil Aviator*, Smithsonian Institution Press, Washington, DC, 1993.

Rhode, Bill; *BALING WIRE, CHEWING GUM, AND GUTS, The Story of the Gates Flying Circus*; Kennikat Press, Port Washington, New York; 1970.

Rose, Robert A., D.D.S.; *LONELY EAGLES, The Story of America's Black Air Force in World War II*, Tuskegee Airmen Inc., Los Angeles Chapter, 1976.

Rose, Robert A., D.D.S.; "Art and the Airman," *Journal , American Aviation Historical Society*, Fall 1974.

Sandler, Stanley; *Segregated Skies*, Smithsonian Institution Press, Washington, D.C. and London.

Schlitz, William P., Senior Editor; "Blacks In U.S. Aviation: THE PIONEERS," *AIR FORCE Magazine*, January 1983.

Schlitz, William P., Senior Editor; "Blacks in U.S. Aviation: FROM TUSKEGEE TO SPACE," *AIR FORCE Magazine*, February, 1983.

"School For Willa," Time Magazine, "National Affairs," September 25, 1939.

Smith, Mary H., "The Incredible Life of Monsieur Bullard," *Ebony Magazine*, New York, December 1967.

Spencer, Chauncey; *Who Is Chauncey Spencer?* Broadside Press, Detroit, MI, 1975.

Taylor, Michael J.H., Editor; *Jane's Encyclopedia of Aviation*, Portland House, New York, 1989.

The Tuskegee Airmen, Official National Publication, Tuskegee Airmen, Inc.

The Tuskegee Airmen, "The Lonely Eagles," *Dispatch*, Confederate Air Force, Volume 12, Number 1, January/February 1987.

Tuskegee Airmen Biographies, Tuskegee Airmen, Inc., Organized in Detroit, MI, 1972.

Tuskegee Airmen Newsletter, Publication of Tuskegee Airmen, Incorporated.

USAF Historical Study No. 85, "USAF Credits for the Destruction of Enemy Aircraft, World War II," Albert F. Simpson Historical Research Center, Air University, Office of Air Force History, Headquarters, USAF, 1978.

Vezina, Meredith R., *In Limbo at Lockett*, The Retired Officer Magazine, February 1994.

Voss, Dorothy; *Forsythe and Anderson Flights*, Tel-News, NJ Bell, January, 1986.

Voss, Dorothy; *New Jersey's Flying Daredevils*, Tel-News, NJ Bell, February, 1989.

Warnock, A. Timothy; *USAF Combat Medals, Streamers, and Campaigns*, USAF Historical Research Center, Officer of USAF History, Washington, DC, 1990.

Warren, James C.; *Colorful Nicknames of TUSKEGEE AIRMEN*, United States of America, Copyright 1993.

332nd Fighter Group, Unit Historical Reports for months of October, 1942 through June, 1945 (missing March, 1944), Albert F. Simpson Historical Research Center, Air University, Maxwell AFB, Alabama.

Index